LEGUMES OF INDO-CHINA

ILDIS
INTERNATIONAL
LEGUME
DATABASE &
INFORMATION
SERVICE

LEGUMES OF INDO-CHINA
A CHECK-LIST

J.M. LOCK

J. HEALD

Published by
Royal Botanic Gardens
Kew

First published 1994

This check-list has been prepared at the Royal Botanic Gardens, Kew, as a contribution to the International Legume Database and Information Service (ILDIS). The aim of the first phase of ILDIS is to produce a computerised database containing basic nomenclatural, distributional and descriptive information on the legumes of the world. At the time of writing, the main data set has been produced by merging an African data set prepared at the Royal Botanic Gardens, Kew, a European data set from the ESFDS Database, and a data set for the Americas provided by the Missouri Botanical Garden. The merged data set is being edited and maintained in content and taxonomic consistency by an international team of experts. The ILDIS project is co-ordinated by Dr F.A.Bisby (University of Southampton), Dr R.M.Polhill (Royal Botanic Gardens, Kew) and Dr J.L.Zarucchi (Missouri Botanical Garden). Institutional sponsors of ILDIS are BIOSIS; Chapman & Hall Scientific Data Division; Royal Botanic Gardens, Kew; V.L. Komarov Botanical Institute, St. Petersburg; National Botanical Research Institute, Lucknow; Missouri Botanical Garden, St. Louis; South China Botanical Institute, Guangzhou, and the University of Southampton. Financial support presently comes from The Leverhulme Trust; IUBS/UNESCO Botany 2000 Programme and the Commonwealth Science Council. Arrangements for accessing the database, purchase of publications and other enquiries may be made to the ILDIS Co-ordinating Centre, Biodiversity & Bioinformatics Research Group, School of Biological Sciences, University of Southampton, Southampton SO9 3TU, U.K.

Cover design by Media Resources, RBG, Kew.

Typeset at Royal Botanic Gardens, Kew by P.Arnold, C.Beard,
D. Costello, M.Newman, H.O'Brien and P.Rosen.

ISBN 0 947643 66 4

Printed and Bound in Great Britain by Whitstable Litho Ltd.,
Whitstable, Kent.

CONTENTS

INTRODUCTION

This check-list of the legumes of Indo-China has been compiled at the Royal Botanic Gardens, Kew, as a contribution to the International Legume Database and Information Service (ILDIS). The data that have been used to prepare this book are held on computer at the Royal Botanic Gardens, Kew, and at the ILDIS Coordinating Centre, whence they are accessible to accredited users.

The area covered is the four countries of Cambodia (Kampuchea), Laos, Thailand and Vietnam. Of the adjoining countries, Peninsular Malaysia is included in Malesia, for which a check-list is currently in preparation at Kew. China will be covered by a separate check-list, being prepared by T-L Wu, T-C Chen & D-X. Zhang at the South China Botanical Institute in Guangzhou (Canton). Burma (Myanma) will be covered as part of South Asia, in which are also included the Andaman and Nicobar Islands in the Bay of Bengal; they are, however, phytogeographically closer to southeast Asia than to Peninsular India.

The check-list has been prepared in the following stages:

1. A search of current Floras of the region, starting with the most recent.
2. Checking of any monographic accounts revealed by (1).
3. A search of the collections in the Herbarium of the Royal Botanic Gardens, Kew, including the folders of cultivated material.
4. Checking of any names which appear in the collections but not in regional Floras or monographs.
5. Checking Index Kewensis and the Kew Record of Taxonomic Literature.
6. Checking of local floristic lists for country records.

The data obtained from these sources were accumulated onto check-sheets, and then entered into an IBM-compatible micro-computer using the program 'ALICE', devised and written by the ALICE Software Partnership (Dr R.Allkin and Mr P.J.Winfield). The completed database contained some 994 taxa (species and infraspecific taxa). The database contains some extra material not included in this printed version, which was produced using the ALICE Report Generator Program (AWRITE). The resulting output was then edited on a word processor and typeset at the Royal Botanic Gardens, Kew.

The preparation of modern Floras for the area under study is actively in progress, although in a somewhat piecemeal fashion. For Thailand, there are recent accounts of both Mimosoideae and Caesalpinioideae in the 'Flora of Thailand'. Coverage of the Papilionoideae is, however, much less complete although there are recent accounts of some genera. The other three countries are being dealt with as a unit in the 'Flore du Cambodge, du Laos et du Viêtnam'. There are recent accounts of Caesalpinioideae and Mimosoideae in this Flora, and also of a number of tribes in the Papilionoideae. The major gaps lie in the Millettieae and Dalbergieae, large and complex tribes urgently in need of attention world-wide. Some work is in progress, particularly on the former. Both Thailand and 'Indo-Chine' were the subject of earlier floras. Craib's 'Florae Siamensis Enumeratio' is little more than a check-list, but Gagnepain's 'Flore Generale de l'Indo-Chine' is a detailed and well-illustrated Flora, although inevitably based on rather few collections, and now much out-of-date. Pham-hoàng Hô's 'An Illustrated Flora of Viet-nam' (Ed. 3, 1991) has not been used. While useful in some ways, its taxonomy is uncritical, and the absence of specimen citation makes the assessment of the names used difficult. It includes some unpublished names.

Information on all the tribes except Phaseoleae was compiled and entered at Kew by Judith Heald during a sandwich year of her course at the University of Bradford. The rest of the material was gathered and entered by J.M. Lock at the Royal Botanic Gardens, Kew.

During the compilation of this check-list both authors received much help and guidance from Dr R.M.Polhill and members of the Legume Section at the Royal Botanic Gardens, Kew. Dr.R. Allkin provided invaluable advice and help with computing matters. Keith Gardner helped with the final editing of the database and the entry of corrections and alterations consequent upon the refereeing process.

Most tribes have been checked by ILDIS taxonomic co-ordinators. Our thanks go to the following: G.P. Lewis (Caesalpinieae, Robinieae), K. Larsen (Cassieae, Cercideae), I. Nielsen (Acacieae, Ingeae, Mimoseae), B. Verdcourt (Abreae), V. Rudd (Aeschynomeneae), L.J.G. van der Maesen (Phaseoleae), R.M. Polhill (Crotalarieae), D. Podlech (Galegeae), B.D.Schrire (Indigofereae, Millettieae), R. Maréchal (Phaseoleae), C.H. Stirton (Psoraleeae, Sophoreae), D. Heller (Trifolieae), H. Fortune Hopkins (Parkieae), H. Ohashi (Desmodieae, Euchresteae) and F. Adema and others at Leiden (Millettieae). Finally, J.E. Vidal (Paris) very kindly checked through the entire database and made numerous useful suggestions. The responsibility for the content of the database lies, however, with J.M. Lock.

THE DATA CATEGORIES

The categories of data in the database are those defined for ILDIS by Dr S.Hollis (1990): Type One data fields: detailed specification. Version 9. (ILDIS, University of Southampton). Not all of these data types are included in this printed checklist.

NOMENCLATURE

Accepted Names

These are species names accepted in one or more standard floras within the region. Such names are only rejected if more recent relevant monographic treatments show this to be necessary.

Subspecies are included in this check-list and given a full entry. The data entered under a species should be the summation of the data entered under all of its subspecies. Varieties are also included in the check-list in full, unless they are varieties of subspecies, when they are mentioned in the 'Notes' section (see below).

There are standard floras for all the countries in the region, but none is complete. For tribes for which there is no standard flora, names have been obtained from all possible sources. Some such sources are old, and taxa only recorded therein cannot have the same level of certainty as those listed in the standard works. Every accepted name in the list is accompanied by the number of a reference (see Bibliography) in which the name is accepted.

A few names are marked as 'Provisional'. Sometimes there is no recent information, and there may be doubt as to the distinctness of the taxon. Likewise listed under this category are taxa which are listed in standard floras but still undescribed (e.g. 'Species A').

In the vast majority of cases the generic names are as agreed by the ILDIS taxonomic co-ordinators.

Synonyms

These are names which have been in use either in standard works or in herbaria since 1940, and which are treated as synonyms of an accepted name in the most recent standard works, or in recent generic revisions. The checklist does not purport to provide an exhaustive synonymy. However, an attempt has been made to account for all names used in Craib's Florae Siamensis Enumeratio and in Gagnepain's account of the family in Lecomte's Flore Générale de l'Indo-Chine. Most basionyms are also included.

Names provisionally placed in synonymy pending discovery of lost types, or because of lack of information, are marked as 'Suspected Synonym'.

Misapplied Names

These are names which have been applied to a taxon which does not include the type of the name. Misapplied names are marked as such in the database, but in this list they may be distinguished by the word *sensu* before the authority, as in (for example) *sensu Craib*, or *sensu auct.*. In the former case the name has been misapplied by one author; in the latter, by more than one.

CHARACTERISTICS

Four categories of basic information are included here. First, is the plant a herb, a shrub or a tree? Herbs are defined as plants producing stems which do not persist from one year to the next. Herbaceous climbers are included here, as are suffrutices, by which is meant plants with a perennial and sometimes woody underground base from which annual herbaceous shoots arise. Few plants of this region fall into this life-form category, although parts of the region with a seasonal climate bear grassland which is regularly burned. Shrubs are woody plants which are branched at or near the base. Lianas (woody climbers) are treated as climbing shrubs. Intermediates are marked as 'Herb or Shrub' or 'Shrub or Tree'.

Secondly, does the plant climb? Those that do, whether by twining, by tendrils, or by any other means, are so marked. Those that may or may not climb, depending on their situation, are marked 'Climbing or not'.

Thirdly, what is the normal lifespan of the plant? Does it persist from one year to the next ('Perennial'), or does it complete its life cycle within a single year and persist to the next as seeds ('Annual')? Once again, variable taxa are marked as such ('Annual or Perennial').

In a few cases there is a note of the conservation status of the plant in the wild, following the conventions applied by the International Union for the Conservation of Nature and Natural Resources (IUCN). Their categories are: Extinct, Endangered, Vulnerable, Indeterminate, Insufficiently Known, and Not Threatened. In the region dealt with here there is generally too little information, but a few taxa have been noted as 'Endangered'.

ECONOMIC IMPORTANCE TO MAN

There is no entry for this category unless there is a recorded use. The classes that have been used are:

Chemical Products includes: All chemicals used in industry, including gums.

Domestic includes: Water purifiers / clarifiers; soap and soap substitutes; cosmetics; chewing sticks and tooth cleaners; hunting gear; brushes; thatching.

Environmental includes: Hedging; ornamental; green manure; shelter; shade; cover crop.

Fibre includes: Cordage; textiles; cork; fibre filling.

Food or drink includes: Vegetables and fruits; beverages; cooking fats and oils; vitamins.
Forage includes: Grazing; browse; fodder; bee plants; host plants.
Medicine includes: Medical and veterinary products.
Miscellaneous includes: Any economic use not otherwise covered. (Such entries are usually accompanied by a note.)
Toxins includes: Materials toxic to any type of living organism.
Weed includes: Agricultural, aquatic, forestry and garden weeds.
Wood includes: Fuel; sawn timber; pole timber; carpentry and construction timber.

Most of these categories are self-explanatory. They represent the top level of a hierarchical classification of plant uses developed at Kew by Mrs Frances Cook (Economic & Conservation Section), which has been adopted as a standard by the International Working Group on Taxonomic Databases for Plant Sciences.

HABITAT AND PHYTOGEOGRAPHY

In 'Legumes of Africa - a Check-list' (Lock 1989, ref. 170) each taxon was given a code corresponding to the phytochorion and the major vegetation types in which it occurs. It has not been found possible to do this adequately for the legumes of Indo-China, mainly because neither of the compilers has field experience in the area.

Takhtajan (1986) placed the whole of the area here considered in his Palaeotropical Kingdom, in the centre of his Indochinese Region. This also includes the most tropical parts of India, much of Burma, the Andaman Islands, and the most tropical parts of southeast China, including Hainan. It should be said that of the three genera that Takhtajan regards as endemic, one (*Endomallus*) is now placed as a synonym of *Atylosia* (q.v.).

The most accessible general account of the vegetation of the whole region appears to be that of Schnell (1962, ref. 171), who also lists many earlier papers. The vegetation of the region is often thought of as dominated by evergreen rain forest, but this is very far from true. There are areas of rain forest, some of them lowland and some montane, but almost the whole area has a marked dry season two months or more long. Much of the region has a tropical seasonal climate and there are large areas of what would in Africa be called 'woodland' (French: forêt claire). Such areas have an open tree layer and a grass understorey which burns regularly. There are also areas of montane grassland, as well as extensive swamps, both freshwater and also estuarine with mangrove.

GEOGRAPHY

The countries included under 'Indo-China' are Cambodia, Laos, Thailand and Viêtnam. The term 'Indo-China' is used solely for convenience and has no political connotations.

Records from adjacent areas are also listed, under the heading 'Asia'. A serious attempt has been made to include all records from contiguous countries — Malaysia, Burma and China. Records for Indonesia, India, Bangladesh and Nepal are also included as often as possible. Records for other countries are sometimes included but should not be regarded as exhaustive. Records for Indonesia and for Malaysia bear the suffix 'ISO' indicating that the whole political country as understood by the International Standards Organisation is

meant.
Status within countries is indicated as Native (N), Introduced by man (I) or of Uncertain status (U).

LITERATURE POINTERS

This section lists references in which may be found a good description, a good illustration, or a distribution map of the whole range of the taxon. Up to three references are listed in each category; the numbers refer to the list at the end of the book.

References to descriptions do not necessarily include the original description; this is normally only included where nothing else is available. Most of the references are to descriptions in English or French.

In selecting illustrations, priority has been given to those showing the whole plant. Most are line drawings.

Only maps showing the complete distribution of a taxon have been included.

VERNACULAR NAMES

The convention within ILDIS is that the only vernacular names included are those which have achieved some international currency. There are very few in this category. 'Flore du Cambodge, du Laos et du Viêt-nam' lists many local vernacular names with their language of origin.

NOTES

Notes that were entered into the computer database have generally been included, although sometimes in an edited form. They may explain entries in other sections, such as the reason for regarding a name as provisional, or the number of a herbarium specimen used as an information source.

COMPILERS' NOTE

This check-list claims to include all taxa recognized in works received at Kew by the end of 1993. There will, of course, be errors and omissions. We would be most grateful if J.M.Lock could be informed of these, since this material will form part of the ILDIS World Legume Database, which will be continually updated to take account of new taxa, new country records, and other material.

The ILDIS Coordinating Centre can be contacted at: Biology Department, University of Southampton, Southampton SO9 3TU, England. In-house electronic copies of the dataset are available from the Co-ordinating Centre, and the database will soon be available on-line as 'legumeline' through the Bath Information and Data Services (BIDS).

A main ILDIS Database currently contains data for Europe, Africa and the Americas. Data sets for West Asia and the Indian Ocean and Madagascar are also complete, and work is proceeding on data sets for Northern Eurasia, Malesia, China and South Asia.

CAESALPINIOIDEAE

AMHERSTIEAE

AMHERSTIA Wall.

A genus of a single tree species, often regarded as native only in Burma, much prized for its large and attractive flowers.

A. nobilis Wall. [17]
Not climbing [17]; Tree [17]; Perennial [17].
Indo-China: Thailand(U) [17]. Asia: Burma(N) [17].
Description [17].
Environmental [17].
Cultivated in Thailand. No wild collections but reported as native [17].

TAMARINDUS L.

A genus of a single tree species, widely planted in the Old World tropics for its sweet-sour fruit-pulp. It may be native in southern Madagascar, but could also have originated in tropical Africa or southern India.

T. indica L. [17]
Indo-China: Cambodia(I) [16]; Laos(I) [16]; Thailand(I) [17]; Vietnam(I) [16].
Not climbing [17]; Tree [17]; Perennial [17].
Description [16,17]; Illustration [16,17].
Food or Drink [17]; Medicine [17]; Miscellaneous [17]; Wood [17].
Cultivated in Thailand and all over the tropics for its edible fruit-pulp.

CAESALPINIEAE

ACROCARPUS Wight & Arn.

A genus of two species, both trees, from India and southeast Asia. One species is widely planted in the tropics for timber and as a shade tree.

Acrocarpus fraxinifolius Arn. [16]
Not climbing[16]; Tree[16]; Perennial[16].
Indo-China: Laos(N) [16]; Thailand(N) [16]. Asia: Burma(N) [16]; India(N) [16]; Indonesia-ISO(N) [16].
Description[16,17]; Illustration[16,17].
Wood[17].

CAESALPINIA L.

A large and complex genus of about 150 species, all trees, shrubs or lianas, many climbing or scrambling. The genus is almost certainly polyphyletic and current studies will probably lead to the reinstatement of a number of segregates. Pantropical; some species have been planted as ornamentals.

C. andamanica (Prain) Hattink [16]
Mezoneuron andamanicum Prain [16]; *Mezoneuron kunstleri* Prain [16].
Climbing[16]; Shrub[16]; Perennial[16].
Indo-China: Thailand(N) [16]; Vietnam(N) [16]. Asia: Indonesia-ISO(N) [16]; Malaysia-ISO(N) [16]. Indian Ocean: Andaman Is(N) [16].
Description[16,17]; Illustration[16,17].

C. bonduc (L.)Roxb. [16]
C. bonducella (L.)Fleming [16]; *C. crista* L., p.p. [16]; *Guilandina bonduc* L. [16]; *Guilandina bonducella* L. [16].
Climbing[16]; Shrub[16]; Perennial[16].
Indo-China: Cambodia(N) [16]; Laos(N) [16]; Thailand(N) [17]; Vietnam(N) [16]. Asia: Burma(N) [16]; China(N) [16]; Hong Kong(N) [16]; India(N) [16]; Indonesia-ISO(N) [16]; Malaysia-ISO(N) [16]; Nepal(N) [16]; Sri Lanka(N) [16]; Taiwan(N) [16].
Description[16,17]; Illustration[16,17].
Environmental[16]; Medicine[17].

C. crista L. [16]
C. nuga (L.)Aiton f. [16]; *Guilandina nuga* L. [16].
Climbing or not[16]; Shrub[16]; Perennial[16].
Indo-China: Cambodia(N) [16]; Thailand(N) [17]; Vietnam(N) [16]. Asia: Burma(N) [16]; China(N) [16]; India(N) [16]; Indonesia-ISO(N) [16]; Malaysia-ISO(N) [16]; Sri Lanka(N) [16]; Taiwan(N) [16]. Australasia: Australia(N) [16]. Indian Ocean: Andaman Is(N) [16].
Description[16,17]; Illustration[16,17].

C. cucullata Roxb. [16]
Mezoneurum cucullatum (Roxb.)Wight & Arn. [16]; *Mezoneurum cucullatum* var. *robustum* Craib [16].
Climbing[16]; Shrub[16]; Perennial[16].
Indo-China: Thailand(N) [16]; Vietnam(N) [16]. Asia: Burma(N) [16]; China(N) [16]; India(N) [16]; Indonesia-ISO(N) [16] Malaysia-ISO(N) [16]; Nepal(N) [16].
Description[16,17]; Illustration[16,17].

C. decapetala (Roth)Alston [16]
C. decapetala var. *japonica* (Sieb. & Zucc.)Ohashi [16]; *C. japonica* Sieb. & Zucc. [16]; *C. sepiaria* Roxb. [16]; *C. sepiaria* var. *japonica* (Sieb. & Zucc.)Gagnep. [16]; *Reichardia decapetala* Roth [16].
Climbing or not[16]; Shrub[16]; Perennial[16].
Indo-China: Laos(N) [17]; Thailand(N) [17]; Vietnam(N) [17]. Asia: Bhutan(N) [17]; Burma(N) [17]; China(N) [17]; India(N) [17] Indonesia-ISO(N) [17]; Japan(N) [17]; Malaysia-ISO(N) [17]; Nepal(N) [17]; Sri Lanka(N) [17].
Description[16,17]; Illustration[16,17].
Cultivated and sometimes naturalised in tropical Africa and America [16].

C. digyna Rottler [16]
Climbing or not[16]; Shrub[16]; Perennial[16].
Indo-China: Cambodia(N) [16]; Laos(N) [16]; Thailand(N) [16]; Vietnam(N) [16]. Asia: Burma(N) [16]; China(N) [16]; India(N) [16]; Indonesia-ISO(N) [16] Malaysia-ISO(N) [16]; Nepal(N) [16]; Sri Lanka(N) [16].
Description[16,17]; Illustration[16,17].
Chemical Products[16]; Domestic[16]; Environmental[16]; Food or Drink[16]; Medicine[16].
Marcan 351: 'Tree 10 ft. high' [4].

C. enneaphylla Roxb. [16]
Mezoneuron enneaphyllum (Roxb.)Benth. [16].
Climbing[16]; Shrub[16]; Perennial[16].
Indo-China: Thailand(N) [16]; Vietnam(N) [16]. Asia: Burma(N) [16]; China(N) [16]; India(N) [16]; Indonesia-ISO(N) [16]; Malaysia-ISO(N) [16].
Description[16,17]; Illustration[16,17].

C. furfuracea (Prain)Hattink [17]
Mezoneuron furfuraceum Prain [17].
Indo-China: Thailand(N) [17]. Asia: Burma(N) [17]; Indonesia-ISO(N) [17].
Climbing[17]; Shrub[17]; Perennial[17].
Description[17,18]; Illustration[17,18].

C. godefroyana Kuntze [16]

C. thorelii Gagnep. [16].
Climbing or not[16]; Shrub[16]; Perennial[16].
Asia: Indo-China: Cambodia(N) [16]; Thailand(N) [16]; Vietnam(N) [16].
Description[16,17]; Illustration[16,17].
Medicine[16].

C. hymenocarpa (Prain)Hattink [16]

Mezoneuron hymenocarpum Prain [16]; *Mezoneuron laoticum* Gagnep. [16].
Indo-China: Cambodia(N) [16]; Laos(N) [16]; Thailand(N) [16]; Vietnam(N) [16]. Asia: Burma(N) [16]; China(N) [16]; Indonesia-ISO(N) [16]; Sri Lanka(N) [16]; Indian Ocean: Andaman Is(N) [16].
Climbing[16]; Shrub[16]; Perennial[16]
Description[16,17]; Illustration[16,17].
Food or Drink[4]; Medicine[4].
Collins 1351[4].

C. latisiliqua (Cav.)Hattink [16]

Mezoneurum balansae Gagnep. [16]; *Mezoneuron keo* Gagnep. [16]; *Mezoneurum latisiliquum* (Cav.)Merr. [16]; *Mezoneurum oxyphyllum* Gagnep. [16].
Climbing[16]; Shrub[16]; Perennial[16].
Indo-China: Vietnam(N) [16]. Asia: Malaysia-ISO(N) [16]; Philippines(N) [16].
Description[16]; Illustration[16].

C. major (Medikus)Dandy & Exell [16]

C. bonduc sensu auctt. [16]; *C. jayabo* Maza, p.p. [16]; *Bonduc major* Medik. [16].
Indo-China: Cambodia(N) [16]; Thailand(N) [16]; Vietnam(N) [16]. Asia: Burma(N) [16]; India(N) [16]; Malaysia-ISO(N) [16]; Sri Lanka(N) [16]. Indian Ocean: Madagascar(N) [16].
Climbing[16]; Shrub[16]; Perennial[16].
Description[16,17]; Illustration[16,17].
Food or Drink[16]; Medicine[16].

C. mimosoides Lam. [16]

Climbing or not[16]; Shrub or Tree[16]; Perennial[16].
Indo-China: Laos(N) [16]; Thailand(N) [16]; Vietnam(N) [16]. Asia: Burma(N) [16]; China(N) [16]; India(N) [16].
Description[16,17]; Illustration[16,17].
Food or Drink[17].

C. minax Hance [16]

C. minax var. *burmanica* Prain [16]; *C. morsei* Dunn [16].
Indo-China: Laos(N) [16]; Thailand(N) [16]; Vietnam(N) [16]. Asia: Burma(N) [16]; China(N) [16]; Hong Kong(N) [16]; India(N) [16] Taiwan(N) [16].
Climbing[16]; Shrub[16]; Perennial[16].
Description[16,17]; Illustration[16,17].

C. nhatrangensis (Gagnep.)J.E.Vidal [16]

Mezoneuron nhatrangense Gagnep. [16].
Climbing[16]; Shrub[16]; Perennial[16].
Indo-China: Vietnam(N) [16].
Description[16]; Illustration[16].
Apparently endemic to southern Vietnam [16].

C. parviflora Prain [17]

C. macra Craib [17].
Climbing or not[17]; Shrub or Tree[17]; Perennial[17].
Indo-China: Thailand(N) [17]. Asia: Indonesia-ISO(N) [17]; Malaysia-ISO(N) [17],Borneo[17] Philippines(N) [17].
Description[17]; Illustration[17].

C. pubescens (Desf.)Hattink [16]

Mezoneuron pubescens Desf. [16]; *Mezoneuron pubescens* var. *longipes* Craib [16].
Climbing[16]; Shrub[16]; Perennial[16].
Indo-China: Thailand(N) [16]; Vietnam(N) [16]. Asia: India(N) [16]; Indonesia-ISO(N) [16].
Description[16,17]; Illustration[16,17].

C. pulcherrima (L.)Sw. [16]
Poinciana pulcherrima L. [16].
Not climbing[17]; Shrub or Tree[17]; Perennial[17].
Indo-China: Cambodia(I) [16]; Laos(I) [16]; Thailand(I) [17]; Vietnam(I) [16]. Asia: Burma(I) [16]; China(I) [16]; Hong Kong(I) [16]; India(I) [16]; Indonesia-ISO(I) [16]; Japan(I) [16]; Malaysia-ISO(I) [16]; Nepal(I) [16]; Sri Lanka(I) [16]; Taiwan(I) [16].
Description[16,17]; Illustration[16,17].
Environmental[17]; Medicine[17].
Peacock Flower[17].
Probably originating from Mexico and/or Guatemala.
Cultivated and naturalised throughout the tropics [17].
Several colour forms.

C. rhombifolia J.E.Vidal [16]
Not climbing[16]; Shrub[16]; Perennial[16].
Indo-China: Vietnam(N) [16].
Description[16]; Illustration[16].
Species known only from the area of the type [16].

C. sappan L. [16]
Not climbing[16]; Shrub or Tree[17]; Perennial[16].
Indo-China: Cambodia(N) [16]; Laos(N) [16]; Thailand(N) [16]; Vietnam(N) [16]. Asia: Burma(N) [16]; China(N) [16]; Hong Kong(N) [16]; India(N) [16] Indonesia-ISO(N) [16]; Malaysia-ISO(N) [16]; Sri Lanka(N) [16].
Description[16,17]; Illustration[16,17].
Environmental[17]; Fibre[16]; Medicine[16].
Sappan Wood[17].
Probably the original Brazil wood of commerce, used for its red dye (G.P.Lewis, pers.comm.).

C. sinensis (Hemsley)J.E Vidal [16]
C. stenoptera Merr. [16]; *C. tsoongii* Merr. [16]; *Mezoneuron sinense* Forbes & Hemsley [16].
Climbing[16]; Shrub[16]; Perennial[16].
Indo-China: Laos(N) [16]; Vietnam(N) [16]. Asia: Burma(N) [16]; China(N) [16]; Hong Kong(N) [16].
Description[16]; Illustration[16].

DELONIX Raf.

A small genus of c. 10 species, most in Madagascar. All are trees and some have showy flowers and are widely planted as ornamentals.

D. elata (L.)Gamble [16]
Poinciana elata L. [19].
Not climbing[19]; Tree[19]; Perennial[19].
Description[19].
Cultivated for the beauty of its flowers [19].
Noted as cultivated in Gagnepain's Flore Generale de l'Indo-Chine [16]. No localised specimens seen.

D. regia (Hook.)Raf. [16]
Poinciana regia Hook. [16].
Not climbing[16]; Tree[16]; Perennial[16].
Indo-China: Cambodia(I) [16]; Laos(I) [16]; Thailand(I) [16]; Vietnam(I) [16]. Indian Ocean: Madagascar(N) [16].
Description[16,17].
Environmental[16].
Flamboyant[16]; Flame Tree[16].
Cultivated throughout the tropics for the beauty and abundance of the large red or, (rarely) orange or yellow, flowers [16].
One of the most common ornamental trees in Thailand [17].
Native of Madagascar.

ERYTHROPHLEUM R.Br.

Nine species, all trees, confined to the Old World tropics. Some species yield timber and most, if not all, contain toxic alkaloids with a variety of recorded uses.

E. fordii Oliver [16]
Not climbing[16]; Tree[16]; Perennial[16].
Indo-China: Vietnam(N) [16]. Asia: China(N) [16].
Description[16]; Illustration[16].
Environmental[16]; Medicine[16]; Toxins[16]; Wood[16].
Conservation status is 'Vulnerable' [16].

E. succirubrum Gagnep. [16]
E. teysmannii var. *puberulum* Craib [17].
Not climbing[16]; Tree[16]; Perennial[16].
Indo-China: Cambodia(N) [16]; Thailand(N) [16].
Description[16,17]; Illustration[16,17].
Toxins[16]; Wood[16].
Probably exists in central Laos. Conservation status is 'Vulnerable' [16].

E. teysmannii (Kurz)Craib [16]
Albizia cambodiana Pierre [16]; *Albizia teysmannii* Kurz [16]; *Entada treas* Gagnep. [16]; *Erythrophleum cambodianum* (Pierre)Gagnep. [16].
Not climbing[16]; Tree[16]; Perennial[16].
Indo-China: Cambodia(N) [16]; Thailand(N) [16].
Description[16,17]; Illustration[16,17].
Medicine[16]; Wood[16].
One of the best timbers in Thailand [16].

GLEDITSIA L.

A small genus of trees and shrubs, mostly spiny, widely dispersed throughout the warm temperate regions but most diverse in eastern Asia. Some yield timber; others are ornamental.

G. australis Hemsley [16, 147]
G. microcarpa Metcalf [147]; *Pogocybe entadoides* Pierre [16].
Not climbing[16]; Tree[16]; Perennial[16].
Indo-China: Vietnam(N) [16]. Asia: China(N) [16].
Description[16]; Illustration[16,147].
Medicine[16].
Often cultivated in Vietnam [16].

G. fera (Lour.)Merr. [16,20]
G. formosana Hayata [20]; *G. rolfei* S.Vidal [16,20]; *G. thorelii* Gagnep. [20]; *Mimosa fera* Lour. [16].
Not climbing[20]; Tree[20]; Perennial[20].
Indo-China: Laos(N) [16]; Thailand(N) [20]; Vietnam(N) [16]. Asia: China(N) [16]; Philippines(N) [16]; Taiwan(N) [16].
Description[16,20]; Illustration[16].
Medicine[20].
G. rolfei S.Vidal is not mentioned by Chen Te-chao (147). Gordon (1966 — unpublished thesis, Indiana University) regards *G. rolfei* as a good species, with *G. formosana* and *G. thorelii* in synonymy.

G. pachycarpa Gagnep. [16]
Indo-China: Vietnam(N) [16].
Not climbing[16]; Tree[16]; Perennial[16].
Description[16].
Only known from the type; in montane forest [16].
Gordon (op.cit.) considers this a synonym of *G. fera*.

GYMNOCLADUS Lam.

A genus of five woody species, one in N America, the rest in E Asia.

G. angustifolius (Gagnep.)J.E.Vidal [16]

Erythrophleum angustifolium Gagnep. [16].
Not climbing[16]; Tree[16]; Perennial[16].
Indo-China: Vietnam(N) [16].
Description[16]; Illustration[16].
Endemic to Northern Vietnam. Known only from 2 collections. Close to *G. burmanicus* C.E.Parkins. [16].

PARKINSONIA L.

Small spiny trees, mostly American but with four species in Africa. One American species is extensively planted in dry tropical and sub-tropical areas for shade, ornament and fuel.

P. aculeata L. [16]

Not climbing[16]; Shrub or Tree[16]; Perennial[16].
Indo-China: Cambodia(I) [16]; Laos(I) [16]; Thailand(I) [17]; Vietnam(I) [16].
Description[16,17].
Jerusalem Thorn[17].
Sometimes cultivated in Indochina. Originates from tropical Central America [16].

PELTOPHORUM (Vogel)Walp.

Trees; about eight species scattered through the tropics. Several species widely cultivated for shade and for their showy yellow flowers.

P. dasyrrhachis (Miq.)Kurz [16]

Not climbing[16]; Tree[16]; Perennial[16].
Indo-China: Cambodia(N) [17]; Laos(N) [17]; Thailand(N) [17]; Vietnam(N) [17] Asia: Indonesia-ISO(N) [17]; Malaysia-ISO(N) [17].
Description[16,17]; Illustration[16,17].
Environmental[17]; Medicine[17]; Wood[17].
Sometimes cultivated in Thailand but not as often as *P. pterocarpum*. Widely cultivated in other countries [17].

var. **dasyrrhachis** [16]

Baryxylum dasyrrhachis Pierre [16]; *Caesalpinia dasyrrhachis* Miq. [16].
Not climbing[16]; Tree[16]; Perennial[16].
Indo-China: Cambodia(N) [16]; Laos(N) [16]; Thailand(N) [16]; Vietnam(N) [16]. Asia: Indonesia-ISO(N) [16]; Malaysia-ISO(N) [16].
Description[16]; Illustration[16].
Environmental[16]; Medicine[16]; Wood[16].

var. **tonkinensis** (Pierre)K.& S.Larsen [16]

Baryxylum tonkinense Pierre [16]; *Peltophorum tonkinense* (Pierre)Gagnep. [16].
Indo-China: Vietnam(N) [16].
Not climbing[16]; Tree[16]; Perennial[16].
Description[16]; Illustration[16].

P. pterocarpum (DC.)K.Heyne [16]

Baryxylum inerme (Roxb.)Pierre [16]; *Inga pterocarpa* DC. [16]; *Peltophorum ferrugineum* (Decne.)Benth. [16]; *Peltophorum inerme* (Roxb.)Naves & Villar [16].
Not climbing[16]; Tree[16]; Perennial[16].
Indo-China: Cambodia(N) [17]; Thailand(N) [16]; Vietnam(N) [16]. Asia: Indonesia-ISO(N) [16]; Malaysia-ISO(N) [16]; Sri Lanka(N) [16]. Australasia: Australia(N) [16].
Description[16,17]; Illustration[16,17].
Chemical Products[16]; Environmental[16]; Medicine[4]; Wood[16].
Luang Vanpruk 286 [4].
Widely introduced into tropical areas. Cultivated in Thailand [17].
Often confused with *P. dasyrrhachis* [16].

PTEROLOBIUM Wight & Arn.

A small genus of about 10 species of spiny climbers from the tropical and sub-tropical parts of the Old World.

P. integrum Craib [16]

P. micranthum sensu auctt. [16].
Climbing or not[16]; Shrub or Tree[16]; Perennial[16].
Indo-China: Cambodia(N) [16]; Laos(N) [16]; Thailand(N) [16]; Vietnam(N) [16].
Description[16,17,171]; Distribution Map [171]; Illustration[16,17,171].

P. macropterum Kurz [16]

P. indicum var. *macropterum* (Kurz)Bak. [16].
Climbing[16]; Shrub[16]; Perennial[16].
Indo-China: Laos(N) [16]; Thailand(N) [16]; Vietnam(N) [16]. Asia: Burma(N) [16]; Indonesia-ISO(N) [16]. Indian Ocean: Andaman Is(N) [16].
Description[16,17,171]; Distribution Map [171]; Illustration[16,17,171].

P. micranthum Gagnep. emend. Craib [17]

Climbing[17]; Shrub[17]; Perennial[17].
Indo-China: Thailand(N) [17].
Description[17,171]; Distribution Map [171]; Illustration[17,171].
Endemic to Thailand [17].

P. microphyllum Miq. [17]

P. densiflorum sensu auct. [17]; *P. indicum* A.Rich. var. *microphyllum* (Miq.)Bak. [16]; *P. insigne* Miq. [171]; *P. platypterum* Gagnep. [16]; *P. punctatum* var. *opacum* Gagnep. [16]; *P. schmidtianum* Harms [16].
Climbing[16]; Shrub[16]; Perennial[16].
Indo-China: Cambodia(N) [16]; Laos(N) [16]; Thailand(N) [16]; Vietnam(N) [16]. Asia: Burma(N) [16]; Indonesia-ISO(N) [16].
Description[16,17,171]; Distribution Map [171]; Illustration[16,17,171].

CASSIEAE

CASSIA L.

The old genus *Cassia* sens. lat. has been split into three in America, Africa and Australia. Not all relevant combinations have, however, been made in the Indo-China region. The members of *Cassia* sens. strict. are mostly trees with more-or-less cylindrical indehiscent fruits, three large sigmoidally-curved filaments and anthers dehiscing by slits.

C. bakeriana Craib [17]

Not climbing[17]; Tree[17]; Perennial[17].
Indo-China: Thailand(N) [17]. Asia: Burma(N) [17].
Description[17].
Medicine[17]; Wood[17].
A hybrid, *C. bakeriana* × *C. fistula* is cultivated under the name Rainbow Shower [17].

C. fistula L. [17]

C. rhombifolia Roxb. [17]; *Bactyrilobium fistulum* (L.)Willd. [17]; *Cathartocarpus fistula* (L.)Pers. [17]; *Cathartocarpus rhombifolius* G.Don [17].
Not climbing[17]; Tree[17]; Perennial[17].
Indo-China: Cambodia(U) [16]; Laos(U) [16]; Thailand(U) [17]; Vietnam(U) [16]. Asia: China(I) [17]; India(N) [17]; Indonesia-ISO(N) [17]; Malaysia-ISO(N) [17]; Sri Lanka(N) [17].
Description[16,17]; Illustration[16,17].
Chemical Products[4]; Food or Drink[16]; Medicine[17]; Wood[17].
Golden Shower[17]; Indian Laburnum[17].; Pudding-pipe Tree[17].
Lakshnakara 1375: 'Bark used for tannin' [4].
Probably native of India, Sri-Lanka and Malesia. At an early date spread to China, Egypt and later throughout the tropics for medical purposes, but mainly as an ornamental [17].

C.grandis L.f. [17]

C. brasiliana Lam. [17]; *C. mollis* Vahl [17]; *C. pachycarpa* De Wit [17].
Not climbing[17]; Tree[17]; Perennial[17].
Indo-China: Laos(I) [16]; Thailand(I) [17]; Vietnam(I) [16].
Description[17]; Illustration[17].
Environmental [17].
Native of tropical America; cultivated throughout the tropics [17].

C. javanica L. [17]

Not climbing[17]; Tree[17]; Perennial[17].
Indo-China: Cambodia(U) [16]; Laos(I) [16]; Thailand(I) [17]; Vietnam(I) [16]. Asia: Burma(N) [17]; India(N) [17]; Indonesia-ISO(N) [17]; Malaysia-ISO(N) [17]; Philippines(N) [17].
Description[16,17]; Illustration[16,17].
Environmental[17]; Medicine[17].

subsp. **javanica** L. [17]

Not climbing[17]; Tree[17]; Perennial[17].
Indo-China: Cambodia(I) [16]; Laos(I) [16]; Thailand(I) [17]; Vietnam(I) [16]. Asia: Indonesia-ISO(N) [17]; Philippines(N) [17].
Description[16,17]; Illustration[16].
Environmental[17]; Medicine[17].
Cultivated in Thailand [17].

subsp. **agnes** (De Wit)K.Larsen [149]

C. agnes (De Wit)Brenan [17]; *C. javanica* var. *agnes* De Wit [17]; *C. javanica* var. *indochinensis* (Gagnep.) K. Larsen [149]; *C. javanica* var. *indochinensis* Gagnep. [149].
Not climbing[17]; Tree[17]; Perennial[17].
Indo-China: Cambodia(N) [16]; Laos(N) [17]; Thailand(N) [17]; Vietnam(N) [17]. Asia: India(N) [17]; Malaysia-ISO(N) [17],Borneo[17].
Description[16,17]; Illustration[16,17].
Environmental[17]; Food or Drink[16]; Medicine[17]; Toxins[16]; Wood[16].
Commonly cultivated throughout tropical Asia [17].

subsp. **nodosa** (Roxb.)K. & S.Larsen [17]

C. nodosa Roxb. [17].
Indo-China: Cambodia(U) [16]; Laos(I) [16]; Thailand(N) [17]; Vietnam(U) [16]. Asia: Burma(N) [17]; India(N) [17]; Indonesia-ISO(N) [17]; Malaysia-ISO(N) [17].
Not climbing[17]; Tree[17]; Perennial[17].
Description[16,17]; Illustration[17].
Environmental[17]; Medicine[17].
Frequently cultivated [17].
Perhaps native in Cambodia and Vietnam [16].

CHAMAECRISTA Moench

A pantropical segregate from the old *Cassia* sens. lat. Most are woody herbs or shrubs. The stamens are subequal and dehisce by apical pores, and the pods are elastically dehiscent. The seeds lack an areole. Some, if not all, are nodulated.

C. absus (L.)Irwin & Barneby [27]

Cassia absus L. [27].
Not climbing[17]; Herb[17], Shrub; Perennial[17].
Indo-China: Thailand(N) [17]; Vietnam(N) [16].
Description[16,17].
Weed[17].
Rare in S.E. Asia although widespread in the tropics [16].

C. leschenaultiana (DC.)Degener [26]

Cassia leschenaultiana DC. [26]; *Cassia mimosoides* var. *wallichiana* (DC.)Bak. [26].
Not climbing[17]; Herb[17], Shrub; Perennial[17].
Indo-China: Cambodia(N) [16]; Laos(N) [16]; Thailand(N) [17]; Vietnam(N) [16].
Description[16,17]; Illustration[16,17].
Irwin & Barneby regard this as synonymous with *Chamaecrista nictitans* (L.)Moench ssp. *patellaria* (Colladon)Irwin & Barneby. Further investigation is needed [27].

C. mimosoides (L.)Greene [26]

Cassia mimosoides L. [17].
Indo-China: Cambodia(N) [16]; Laos(N) [16]; Thailand(N) [17]; Vietnam(N) [16].
Not climbing[17]; Herb[17], Shrub; Perennial[17].
Description[16,17]; Illustration[16,17].
Probably indigenous to continental S.E. Asia. *C. mimosoides* & *C. leschenaultiana* are fairly distinct in Thailand [17].

C. pumila (Lam.) K.Larsen [149]

Cassia pumila Lam. [149]; *Senna prostrata* Roxb. [16].
Not climbing[17]; Herb[17], Shrub; Perennial[17].
Indo-China: Cambodia(N) [16]; Laos(N) [16]; Thailand(N) [17]; Vietnam(N) [16]. Australasia: Australia(N) [17].
Description[16,17]; Illustration[16,17].
Weed[17].
Common all over Thailand, but not as common as *C.mimosoides*. Perhaps *C.kleinii* Wight & Arn. belongs here [17].

DIALIUM L.

A genus of about 40 species, all trees, pantropical. The genus is somewhat heterogeneous but all species share a small indehiscent 1–2-seeded fruit.

D. cochinchinense Pierre [17]

Not climbing[17]; Tree[17]; Perennial[17].
Indo-China: Cambodia(N) [17]; Laos(N) [17]; Thailand(N) [17]; Vietnam(N) [17].
Description[16,17]; Illustration[16,17].
Domestic[16]; Fibre[16]; Food or Drink[16,17]; Medicine[16].
Wood[16,17].
Velvet Tamarind[17].
Closely related to *D. indum* & *D. maingayi* from the Malay Peninsula [17].

D. indum L. [17]

D.javanicum Burm.f. [17].
Not climbing[17]; Tree[17]; Perennial[17].
Indo-China: Thailand(N) [17]. Asia: Indonesia-ISO(N) [17]; Malaysia-ISO(N) [17].
Description[17]; Illustration[17].
Food or Drink[17]; Wood[17].

D. patens Bak. [20]

D. angustisepalum Ridl. [20].
Not climbing[20]; Tree[20]; Perennial[20].
Indo-China: Thailand(N) [20]. Asia: Indonesia-ISO(N) [20]; Malaysia-ISO(N) [20], Borneo[20]; Singapore(N) [20].
Description [20].
Food or Drink [4].
C. Niyomdham et al. 1184 'mature fruits edible' [4].

D. platysepalum Bak. [17]

Not climbing[17]; Tree[17]; Perennial[17].
Indo-China: Thailand(N) [17]. Asia: Indonesia-ISO(N) [17]; Malaysia-ISO(N) [17], Borneo[17].
Description[17].
Food or Drink[17]; Wood[17]

KOOMPASSIA Benth.

Three species, all trees, and all found in southeast Asia and Malesia. *K. excelsa* is the tallest known tropical angiosperm.

K. excelsa (Becc.)Taubert [17]

K. parviflora King [17]; *Abauria excelsa* Becc. [17].
Not climbing[17]; Tree[17]; Perennial[17].
Indo-China: Thailand(N) [17]. Asia: Indonesia-ISO(N) [17]; Malaysia-ISO(N) [17].
Description[17].
Wood[17]

K. malaccensis Benth. [17]

K. beccariana Taubert [17].
Not climbing[17]; Tree[17]; Perennial[17].
Indo-China: Thailand(N) [17]. Asia: Indonesia-ISO(N) [17]; Malaysia-ISO(N) [17].
Description[17].
Wood[17].

SENNA Miller

A large genus, mainly of shrubs and trees. One of the segregates from the old *Cassia* sens.lat. The stamens dehisce by apical pores and are all more-or-less straight, the pods dehisce tardily or not at all, and the seeds often possess an areole. Few (if any) species are nodulated.

S. alata (L.)Roxb. [17,26]

Cassia alata L. [26].
Not climbing[17]; Shrub[17]; Perennial[17].
Indo-China: Cambodia(I) [16]; Laos(I) [16]; Thailand(I) [17]; Vietnam(I) [16]. South America.
Description[16,17]; Illustration[16,17].
Medicine[17].
Candelabra Bush[17]; Ring-worm Bush[17].
Cultivated for medical purposes; sometimes grown in gardens. Native of S.America but now pantropical [16].

S. bacillaris (L.f.)Irwin & Barneby [27]

Cassia bacillaris L.f. [27]; *Cassia fruticosa* sensu auctt. [27].
Indo-China: Thailand(I) [17]. Central America.
Not climbing[17]; Shrub or Tree[17]; Perennial[17].
Description[17].
Environmental[17]; Medicine[17].
Cultivated in Thailand, but not frequent [17]. Irwin & Barneby attribute cultivated material named as *C. fruticosa* to this taxon [27].

S. bicapsularis (L.)Roxb. [17,26]

Cassia bicapsularis L. [26]; *Cassia laevigata* sensu auct. [17].
Not climbing[17]; Shrub[17]; Perennial[17].
Indo-China: Thailand(I) [17]; Vietnam(I) [16]. Asia: India(I) [17]. South America.
Description[16,17]; Illustration[16,17].
Environmental[16,17]; Food or Drink[16]; Wood[16].
Cultivated, but not common in Thailand [17].

S. garrettiana (Craib)Irwin & Barneby [27]

Cassia garrettiana Craib [27].
Not climbing[17]; Tree[17]; Perennial[17].
Indo-China: Cambodia(N) [17]; Laos(N) [17]; Thailand(N) [17]; Vietnam(N) [17].
Description[16,17]; Illustration[16,17].
Environmental[16]; Food or Drink[17]; Medicine[16]; Wood[17].
Endemic to the Indochinese Peninsular [16].
Often planted as a wayside tree - not wild in the peninsular [17].

S. hirsuta (L.)Irwin & Barneby [27]

Cassia hirsuta L. [27].
Not climbing[17]; Herb or Shrub[17]; Perennial[17].
Indo-China: Cambodia(I) [16]; Laos(I) [16]; Thailand(I) [17]; Vietnam(I) [16]. South America.
Description[16,17]; Illustration[16,17].
Food or Drink[16]; Medicine[17]; Weed[17].
Now a weed in many places in the tropics, but not very common [17].

S. occidentalis (L.)Link [17,27]

Cassia occidentalis (L.)Rose [27]; *Ditramexa occidentalis* Britt. & Rose [17].
Not climbing[17]; Herb or Shrub[17]; Perennial[17].
Indo-China: Cambodia(I) [16]; Laos(I) [16]; Thailand(I) [17]; Vietnam(I) [16].
Description[16,17]; Illustration[16,17].
Food or Drink[16,17]; Medicine[16]; Weed[17].
Cafe Negre[16]; Casse Puante[16].

S. septemtrionalis (Viviani)Irwin & Barneby [27]

Cassia floribunda sensu auctt. [16,27]; *Cassia laevigata* Willd. [16].
Indo-China: Vietnam(I) [16].
Not climbing[16]; Herb or Shrub[16]; Perennial[16].
Description[16]; Illustration[16].
Environmental[16].
Originated in S.America. Rare in Indochina [16].

S. siamea (Lam.)Irwin & Barneby [17,27]

Cassia siamea Lam. [27].
Indo-China: Cambodia(N) [16]; Laos(N) [16]; Thailand(N) [17]; Vietnam(N) [16].
Not climbing[17]; Tree[17]; Perennial[17].
Description[16,17]; Illustration[16,17].
Food or Drink[17]; Wood[17].

S. sophera (L.)Roxb. [27]

Cassia sophera L. [27].
Indo-China: Cambodia(U) [16]; Laos(U) [16]; Thailand(U) [17]; Vietnam(U) [16].
Not climbing[17]; Shrub[17]; Perennial[17].
Description[16,17]; Illustration[16,17].
Medicine[17]; Weed[17].
Found throughout the tropics. Related to *C. occidentalis*, but clearly a distinct species [17].

S. spectabilis (DC.)Irwin & Barneby [27]

Cassia spectabilis DC. [27].
Indo-China: Thailand(I) [17].
Not climbing[17]; Tree[17]; Perennial[17].
Description[17]; Illustration[17].
Environmental[17].

S. sulfurea (Colladon)Irwin & Barneby [17]

Cassia glauca Lam. [17]; *Cassia surattensis* sensu auctt. [27]; *Cassia surattensis* Burm. f. subsp. *glauca* (Lam.)K. & S.Larsen [27].
Indo-China: Laos(N) [16]; Thailand(N) [17]; Vietnam(N) [16]. Asia: India(N) [17].
Not climbing[17]; Shrub or Tree[17]; Perennial[17]
Description[16,17]; Illustration[16,17].
Environmental[16]; Food or Drink[16].

S. surattensis (Burm. f.)Irwin & Barneby [27]

Cassia glauca var. *suffruticosa* (Roth)Bak. [27]; *Cassia suffruticosa* Roth [27]; *Cassia surattensis* Burm. f. [27]; *Cassia surattensis* subsp. *suffruticosa* (Heyne)K. & S.Larsen [27].
Not climbing[17]; Herb or Shrub[17]; Perennial[17].
Indo-China: Laos(I) [16]; Thailand(I) [17]; Vietnam(I) [16].
Description[16,17]; Illustration[16,17].
Environmental[16]; Food or Drink[16]; Medicine[16].
Often cultivated in Thailand. Cultivated throughout the tropics [17].

S. timoriensis (DC.)Irwin & Barneby [17,27]

Cassia arayatensis Llanos [17]; *Cassia montana* Naves & Villar [17]; *Cassia timoriensis* DC. [27].
Not climbing[17]; Shrub or Tree[4]; Perennial[17].
Indo-China: Cambodia(N) [16]; Laos(N) [16]; Thailand(N) [17]; Vietnam(N) [16]. Asia: Sri Lanka(N) [17]. Australasia: Australia(N) [17].
Description[16,17]; Illustration[16,17].
Food or Drink[16]; Medicine[16]; Wood[16].
Often seen as a pioneer species. The densely golden hairy form which seems to to be the most common in Thailand, has been recognised (e.g. as forma *xanthocoma* (Miq.)De Wit). As clinal variation from nearly glabrous to densely hairy can be found it does not seem reasonable to maintain a separate name [17].

S. tora (L.)Roxb. [27]

Cassia borneensis Miq. [16]; *Cassia tora* L. [27]; *Cassia tora* var. *borneensis* (Miq.)Miq. [16,17].
Indo-China: Cambodia(N) [16]; Laos(N) [16]; Thailand(N) [17]; Vietnam(N) [16].
Not climbing[17]; Herb or Shrub[17]; Perennial[17].
Description[16,17]; Illustration[16,17].
Medicine[16]; Weed[17].
Common weed throughout Thailand. Pantropic weed [17].

ZENIA Chun

A genus of a single species; poorly known. Tree.

Zenia insignis Chun [16]

Not climbing[16]; Tree[16]; Perennial[16].
Indo-China: Vietnam(N) [16]. Asia: China(N) [16].
Description[16]; Illustration[16,147].
Environmental[16]; Wood[28].

CERCIDEAE

BAUHINIA L.

A very large pantropical genus of trees, shrubs and lianas. Some have large decorative flowers and have been widely planted. The present general view is that the genus should be treated broadly; earlier workers in this region have divided the species among several genera.

B. acuminata L. [17]

B.candida sensu auct. [17]; *B. grandiflora* sensu auct. [17]; *B. linnaei* Ali [16]; *B. tomentosa* sensu auct. [17].
Not climbing[17]; Shrub[17]; Perennial[17].
Indo-China: Cambodia(I)[16]; Laos(I)[16]; Thailand(I)[17]; Vietnam(I)[16]. Asia: Indonesia-ISO(I)[17]; Philippines(I)[17].
Description[16,17]; Illustration[16,17].
Commonly cultivated all over S.E Asia and Sri Lanka. May be found as an escape in deciduous forests and scrub [17].

B. aureifolia K. & S. Larsen [150]

B.chrysophylla K. & S. Larsen [150].
Indo-China: Thailand(N)[4].
Climbing[4]; Endangered[150]; Shrub[4]; Perennial[4].
Niyomdham 1737 [4]. A few specimens only known in an isolated rainforest patch surrounded by rubber plantations (Larsen, pers.comm.).

B. bassacensis Gagnep. [17]

B. detergens Craib [17]; *B. sanitwongsei* Craib [17]; *Phanera bassacensis* (Gagnep.)De Wit [17].
Climbing[17]; Shrub[17]; Perennial[17].
Indo-China: Cambodia(N)[17]; Laos(N)[17]; Thailand(N)[17]; Vietnam(N)[17].
Description[16,17]; Illustration[16,17].

B. bidentata Jack [17]

Climbing[17]; Shrub[17]; Perennial[17].
Indo-China: Thailand(N)[17]. Asia: Indonesia-ISO(N)[17]; Malaysia-ISO(N)[17].
Description[17]; Illustration[17].

subsp. **bicornuta** (Miq.)K. & S.Larsen [17]

B. emarginata Jack [17]; *B. gracilipes* Merr. [17]; *Phanera bicornuta* Miq. [17]; *Phanera bidentata* subsp. *bicornuta* (Miq.)De Wit [17].
Climbing[17]; Shrub[17]; Perennial[17].
Asia: Indonesia-ISO(N)[17]; Malaysia-ISO(N)[17]. Indo-China: Thailand(N)[17].
Description[17]; Illustration[17].
This polymorphic species is divided into several subspecies by de Wit [17].

B. binata Blanco [17]

B. blancoi (Benth.)Bak. [17]; *Lysiphyllum binatum* (Blanco)De Wit [17]; *Phanera blancoi* Benth. [17]; *Phanera complicata* Miq. [17].
Climbing[17]; Shrub[17]; Perennial[17].
Asia: Indonesia-ISO(N)[17]; Philippines(N)[17]; Australasia: Australia(N)[17]. Indo-China: Thailand(N)[17].
Description[17]; Illustration[17].
Perhaps introduced into Thailand [17].

B. brachycarpa Benth. [17]

B. enigmatica Prain [17].
Indo-China: Thailand(N)[17]; Laos(N)[149]. Asia: Burma(N)[17]; China(N)[149].
Not climbing[17]; Shrub or Tree[17]; Perennial[17].
Description[17]; Illustration[17].
In Thailand occurs in a stunted form at 2000 m; c. 50-60cm high with leaves smaller than the type [17].

B. bracteata (Benth.)Bak. [16]

Climbing[16]; Shrub[16]; Perennial[16].
Indo-China: Cambodia(N)[17]; Laos(N)[17]; Thailand(N)[17]; Vietnam(N)[17]. Asia: Burma (N)[17].
Description[16,17]; Illustration[16,17].
Fibre[16]; Food or Drink[4]; Medicine[16].
Collins 1615: 'Flowers eaten as vegetable' [4].

subsp. **bracteata** [16]

B. bracteata var. *marcanii* Craib [16]; *B. harmandiana* Gagnep. [17]; *B. helferi* Craib [17]; *B. nhatrangensis* Gagnep. [16]; *B. unguiculata* Bak. [17]; *Phanera bracteata* Benth. [17].
Climbing[16]; Shrub[16]; Perennial[16].
Indo-China: Cambodia(N)[17]; Laos(N)[17]; Thailand(N) [16]; Vietnam(N)[17]. Asia: Burma (N)[17].
Description[16,17]; Illustration[16,17].
Fibre[16]; Medicine[16].
Collins 1615: 'Flowers eaten as vegetable' [4].

subsp. **astylosa** K. & S.Larsen [16]

Climbing[16]; Shrub[16]; Perennial[16].
Indo-China: Vietnam(N)[16].
Description[16]; Illustration[16].
Fibre[16]; Food or Drink[4]; Medicine[16].

B. calycina Gagnep. [16]

Climbing[16]; Endangered[16]; Shrub[16]; Perennial[16].
Indo-China: Cambodia(N)[16].
Description[16]; Illustration[16].
Endemic to Cambodia; known only from the type locality [16].

B. carcinophylla Merr. [16]

Indo-China: Vietnam(N)[16]. Asia: China(N)[16].
Climbing[16]; Shrub[16]; Perennial[16].
Description[16]; Illustration[16].

B. cardinalis Gagnep. [16]

B. dolichobotrys Merr. [16].
Climbing[16]; Shrub[16]; Perennial[16].
Indo-China: Cambodia(N)[16]; Laos(N)[16]; Vietnam(N)[16].
Description[16]; Illustration[16].

B. championii (Benth.)Benth. [16]

Climbing[16]; Shrub[16]; Perennial[16].
Indo-China: Vietnam(N)[16]. Asia: China(N)[16]; Taiwan(N)[16].
Description[16]; Illustration[16].

var. **championii** [16]

B. bonii Gagnep. [16]; *B. championii* var. *acutifolia* Chen [16]; *B. gnomon* Gagnep. [16]; *B. hunanensis* Hand.-Mazz. [16]; *B. lecomtei* Gagnep. [16]; *B. yingtakensis* Merr. & Metc. [16]; *Phanera championii* Benth. [16].
Climbing[16]; Shrub[16]; Perennial[16].
Indo-China: Vietnam(N)[16]. Asia: China(N)[16]; Taiwan(N)[16].
Description[16]; Illustration[16].

var. **rubiginosa** K. & S.Larsen [16]

Climbing[16]; Shrub[16]; Perennial[16].
Indo-China: Vietnam(N)[16].
Description[16].

B. clemensiorum Merr. [16]
B. clemensorum Gagnep. [16].
Climbing[16]; Endangered[16]; Shrub[16]; Perennial[16].
Indo-China: Vietnam(N)[16].
Description[16]; Illustration[16].
Endemic to central Vietnam [16].

B. coccinea (Lour.)Dc. [16]
Climbing[16]; Shrub[16]; Perennial[16].
Indo-China: Laos(N)[16]; Vietnam(N)[16].
Description[16]; Illustration[16].
Fibre[16].

subsp. **coccinea** [16]
B. mastipoda Gagnep. [16]; *Phanera coccinea* Lour. [16].
Climbing[16]; Shrub[16]; Perennial[16].
Indo-China: Laos(N)[16]; Vietnam(N)[16].
Description[16]; Illustration[16].
Fibre[16].

subsp. **tonkinensis** (Gagnep.)K. & S.Larsen [16]
B. ferruginea var. *tonkinensis* Gagnep. [16].
Climbing[16]; Shrub[16]; Perennial[16].
Indo-China: Vietnam(N)[16].
Description[16].
Endemic to North Vietnam [16].

B. concreta Craib [17]
Climbing or not[17]; Endangered[17]; Shrub[17]; Perennial[17].
Indo-China: Thailand(N)[17].
Description[17]; Illustration[17].
Endemic to Thailand, on limestone rocks [17].

B. corymbosa DC. [16]
Phanera corymbosa (DC.)Benth. [16].
Climbing[16]; Shrub[16]; Perennial[16].
Indo-China: Vietnam(N)[16]. Asia: China(N)[16].
Description[16].

B. curtisii Prain [17]
B. calcicola Craib [17]; *Lasiobema curtisii* (Prain)De Wit [17].
Climbing or not[17]; Shrub[17]; Perennial[17].
Indo-China: Cambodia(N)[17]; Laos(N)[17]; Thailand(N)[17]; Vietnam(N)[17]. Asia: Malaysia-ISO(N)[17].
Description[16,17]; Illustration[16,17].
Fibre[16]; Medicine[16].

B. ferruginea Roxb. [4]
Phanera ferruginea (Roxb.)Benth. [24].
Climbing[4]; Shrub[4]; Perennial[4].
Indo-China: Thailand[16]. Asia: Malaysia-ISO(N)[16].
Description[24].
Niyomdham & W. Veachirakan 1912: 'Thailand, scandent shrub' [4].

B. glabrifolia (Benth.)Bak. [16]
Climbing[16]; Shrub[16]; Perennial[16].
Asia: Burma(N)[16]; Malaysia-ISO(N)[16]. Indo-China: Laos(N)[16]; Thailand(N)[16].
Description[16,17]; Illustration[16,17].

var. **glabrifolia** [25]
Phanera glabrifolia Benth. [17].
Climbing[25]; Shrub[25]; Perennial[25].
Description[25].
Known only from the type sent out from Calcutta Botanic Gardens [25].

var. **maritima** K. & S.Larsen [17]

B. glabrifolia sensu auct. [17]; *B. piperifolia* sensu auct. [17]; *Phanera glabrifolia* sensu auct.[17].
Climbing[17]; Shrub[17]; Perennial[17].
Indo-China: Thailand(N)[17]. Asia: Burma(N)[17]; Malaysia-ISO(N)[17].
Description[17]; Illustration[17].
A rare taxon, endangered in Thailand [149].

var. **sericea** (Lace)K. & S.Larsen [16]

B. glabrifolia subsp. *sericea* (Lace)K. & S.Larsen [16]; *B. prabangensis* Gagnep. [16]; *B. sericea* Lace [16].
Climbing[16]; Shrub[16]; Perennial[16].
Indo-China: Laos(N)[16]; Thailand(N)[16]. Asia: Burma(N)[16].
Description[16,17]; Illustration[16,17].

B. glauca (Benth.)Benth. [16]

Climbing[17]; Shrub[17]; Perennial[17].
Indo-China: Cambodia(N)[17]; Laos(N)[17]; Thailand(N)[17]; Vietnam(N)[17]. Asia: Burma(N)[17]; China(N)[17]; India(N)[17]; Indonesia-ISO(N)[17]; Malaysia-ISO(N)[17].
Description[16,17]; Illustration[16,17].

subsp. **glauca** [17]

B. micrantha Ridley [17]; *B. viridiflora* Miq. [17]; *Phanera glauca* Benth. [17].
Climbing[17]; Shrub[17]; Perennial[17].
Indo-China: Thailand(N)[17]. Asia: Burma(N)[17]; China(N)[17]; India(N)[17]; Indonesia-ISO(N)[17]; Malaysia-ISO(N)[17].
Description[17]; Illustration[17].
Total distribution of this subspecies is not yet satisfactorily elucidated [17].

subsp. **tenuiflora** (C.B.Clarke)K. & S.Larsen [17]

B. caterviflora Chen [17]; *B. polysperma* Gagnep. [17]; *B. tenuiflora* C.B.Clarke [17]; *Phanera tenuiflora* (C.B.Clarke)De Wit [17].
Climbing[17]; Shrub[17]; Perennial[17].
Asia: Burma(N)[17]; China(N)[17]. Indo-China: Cambodia(N)[17]; Laos(N)[17]; Thailand (N)[17]; Vietnam(N)[17].
Description[16,17]; Illustration[16,17].
B.hupehana Craib from SE China is very close to subsp. *tenuiflora*. Further studies of Chinese material are necessary [17].

B. godefroyi Gagnep. [16]

Climbing or not[16]; Endangered[16]; Shrub[16]; Perennial[16].
Indo-China: Cambodia(N)[16].
Description[16]; Illustration[16].
Known only from the type locality. Floral structure of the type appears abnormal [16].

B. harmsiana Hosseus [16]

Climbing[16]; Shrub[16]; Perennial[16].
Indo-China: Cambodia(N)[16]; Thailand(N)[16].
Description[16,17]; Illustration[16,17].

var. **harmsiana** [17]

Climbing[17]; Shrub[17]; Perennial[17].
Indo-China: Cambodia(N)[17]; Thailand(N)[17].
Description[17]; Illustration[17].

var. **media** (Craib)K. & S.Larsen [17]

B. media Craib [17].
Climbing[17]; Shrub[17]; Perennial[17].
Indo-China: Thailand(N)[17].
Description[17].
Only known from the type locality [149].

B. hirsuta J.A.Weinm. [17]

B. acuminata var. *hirsuta Craib* [17]; *B. mollissima* sensu auct. [17]; *B. parvula* Gagnep. [17].
Indo-China: Cambodia(N)[17]; Laos(N)[16]; Thailand(N)[17]; Vietnam(N)[17]. Asia: China(N)[17]; Indonesia-ISO(N)[[17]; Malaysia-ISO(N)[17].
Not climbing[17]; Shrub[17]; Perennial[17].
Description[16,17]; Illustration[16,17].
Fibre[16]; Medicine[16].

B. integrifolia Roxb. [17]
B. flammifera Ridl. [17]; *B. holosericea* Ridl. [17]; *Phanera integrifolia* (Roxb.)Benth. [17].
Indo-China: Thailand(N)[17]. Asia: Malaysia-ISO(N)[17].
Climbing[17]; Shrub[17]; Perennial[17].
Description[17]; Illustration[17].
All Thai specimens are subsp. *integrifolia* [17].

B. involucellata Kurz [17]
B. involucellata var. *jaeckelii* K.Larsen [17]; *Phanera involucellata* (Kurz)De Wit [17].
Climbing[17]; Shrub[17]; Perennial[17].
Indo-China: Thailand(N)[17]. Asia: Burma(N)[17].
Description[17]; Illustration[17].
Fibre[4]; Miscellaneous[4].
Khoon Winit 495: 'Bark used for cordage, leaves used for cigar cover' [4].

B. involucrans Gagnep. [16]
Climbing[16]; Shrub[16]; Perennial[16].
Indo-China: Vietnam(N)[16].
Description[16]; Illustration[16].
Known only from the type locality [16].

B. khasiana Bak. [16]
Climbing[16]; Shrub[16]; Perennial[16].
Indo-China: Laos(N)[16]; Vietnam(N)[16]. Asia: China(N)[16]; India(N)[16].
Description[16]; Illustration[16].
Fibre[16].

subsp. **khasiana** [16]
B. bidentata sensu auct. [16]; *B. howei* Merr. & Chun [16]; *B. pierrei* Gagnep. [16].
Climbing[16]; Shrub[16]; Perennial[16].
Indo-China: Laos(N)[16]; Vietnam(N)[16]. Asia: China(N)[16]; India(N)[16].
Description[16]; Illustration[16].
Fibre[16].

subsp. **polystachya** (Gagnep.)K. & S.Larsen [16]
B. polystachya Gagnep. [16].
Climbing[16]; Shrub[16]; Perennial[16].
Indo-China: Laos(N)[16]; Vietnam(N)[16].
Description[16].

B. lakhonensis Gagnep. [16]
B. sepis Craib [16].
Climbing or not[16]; Shrub[16]; Perennial[16].
Indo-China: Laos(N)[16]; Thailand(N)[17]; Vietnam(N)[16].
Description[16,17]; Illustration[16,17].
Medicine[16].

B. lorantha Gagnep. [16]
Climbing[16]; Endangered[[16]; Shrub[16]; Perennial[16].
Indo-China: Laos(N)[16].
Description[16]; Illustration[16].
Endemic to South Laos [16].

B. malabarica Roxb. [17]
B. acida Korth. [16]; *B.castrata* Hassk. [16]; *B. platyphylla* Miq. [16]; *B. rugulosa* Miq. [16]; *Piliostigma acidum* (Korth.)Benth. [16]; *Piliostigma malabaricum* (Roxb.)Benth. [16]; *Piliostigma malabaricum* var. *acidum* (Korth.)De Wit [16].
Not climbing[16]; Tree[16]; Perennial[16].
Indo-China: Cambodia(N)[16]; Laos(N)[16]; Thailand(N)[16]; Vietnam(N)[16]. Asia: Burma(N)[16]; India(N)[16]; Indonesia-ISO(N)[16]; Philippines(N)[16].
Description[16,17]; Illustration[16,17].
Food or Drink[16]; Medicine[4].
Collins 884 [4].

B. monandra Kurz [17]

B. kappleri Sagot [17]; *B. krugii* Urban [17]; *B. richardiana* Voight [16]; *B. subrotundifolia* sensu auct. [17]; *Caspariopsis monandra* (Kurz)Britton & Rose [17].
Not climbing[17]; Shrub or Tree[17]; Perennial[17].
Indo-China: Thailand(U)[17]; Vietnam(U)[17].
Description[16,17].
Origin unknown, probably neotropical. Now pantropical, cultivated & escaped [17].

B. nervosa (Benth.)Bak. [17]

Phanera nervosa Benth. [17].
Climbing[17]; Shrub[17]; Perennial[17].
Indo-China: Thailand(N)[17]. Asia: Burma(N)[17]; China(N)[17]; India(N)[17].
Description[17]; Illustration[17].

B. ornata Kurz [16]

Climbing[16]; Shrub[16]; Perennial[16].
Indo-China: Laos(N)[17]; Thailand(N)[17]; Vietnam(N)[17]. Asia: Burma(N)[17]; China(N)[17]; India(N)[17].
Description[16,17]; Illustration[16,17].
An extremely polymorphic taxon [17].

var. **ornata** [16]

Climbing[16]; Shrub[16]; Perennial[16].
Asia: Burma(N) [16].
Description[16].

var. **balansae** (Gagnep.)K. & S.Larsen [16]

B. balansae Gagnep. [16]; *B. petelotii* Merr. [16].
Indo-China: Vietnam(N)[16].
Climbing[16]; Endangered[16]; Shrub[16]; Perennial[16].
Description[16].
Endemic to northern Vietnam [16].

var. **burmanica** K. & S.Larsen [17]

Climbing[17]; Shrub[17]; Perennial[17].
Indo-China: Thailand(N)[17]. Asia: Burma(N)[17].
Description[17]; Illustration[17].

var. **kerrii** (Gagnep.)K. & S.Larsen [16]

B. bakeriana S.Larsen [16]; *B. eberhardtii* Gagnep. [16]; *B. kerrii* Gagnep. [16]; *B. kerrii* var. *grandiflora* Craib [16]; *B. rufa* (Benth.)Baker [16]; *Phanera rufa* Benth. [16].
Climbing[16]; Shrub[16]; Perennial[16].
Indo-China: Laos(N)[16]; Thailand(N)[17]; Vietnam(N)[16]. Asia: Burma(N)[17]; China(N)[17].
Description[16,17]; Illustration[16,17].

var. **laotica** K. & S.Larsen [4]

Climbing[4]; Shrub[4]; Perennial[4].
Indo-China: Laos(N)[4].
Talbot de Malahede 152[4].

var. **subumbellata** (Gagnep.)K. & S.Larsen [16]

B. inflexilobata Merr. [16]; *B. subumbellata* Gagnep. [16].
Climbing[16]; Shrub[16]; Perennial[16].
Indo-China: Laos(N)[16]; Vietnam(N)[16].
Description[16].

B. oxysepala Gagnep. [16]

Climbing[16]; Endangered[16]; Shrub[16]; Perennial[16].
Indo-China: Vietnam(N)[16].
Description[16]; Illustration[16].
Endemic to northern Vietnam [16].

B. penicilliloba Gagnep. [16]

B. penicilliloba var. *harmandiana* Gagnep. [16].
Climbing or not[16]; Shrub[16]; Perennial[16].
Indo-China: Cambodia(N)[17]; Laos(N)[16]; Thailand(N)[16]; Vietnam(N)[16].
Description[16,17]; Illustration[16,17].
Medicine[4].
R.F.D.'s collector 9003: 'Medicinal roots' [4].
If the ground is burnt annually it rarely becomes a climbing shrub, but produces each year some slim shoots 25–50 cm. high [17].

B. pottsii G.Don [17]

Not climbing[17]; Shrub or Tree[17]; Perennial[17].
Indo-China: Cambodia(N)[17]; Thailand(N)[17]. Asia: Burma(N)[17]; Indonesia-ISO(N)[17]; Malaysia-ISO(N)[17].
Description[16,17]; Illustration[16,17].
Miscellaneous[16].

var. **pottsii** [17]

B. elongata Korth. [17]; *B. pottsii* var. *elongata* (Korth.)De Wit [17]; *Phanera elongata* (Korth.)Benth. [17]; *Phanera speciosa* Miq. [17].
Not climbing[17]; Shrub or Tree[17]; Perennial[17].
Indo-China: Thailand(N)[17]. Asia: Burma(N)[17]; Indonesia-ISO(N)[17]; Malaysia-ISO(N)[17].
Description[17]; Illustration[17].

var. **decipiens** (Craib)K. & S.Larsen [17]

B. decipiens Craib [17].
Not climbing[17]; Shrub or Tree[17]; Perennial[17].
Indo-China: Thailand(N)[17].
Description[17].
Close to var. *subsessilis* [17].

var. **mollissima** (Prain)K. & S.Larsen [17].

B. mollissima Prain [17].
Climbing or not[4]; Shrub or Tree[17]; Perennial[17].
Indo-China: Thailand(N)[17]. Asia: Malaysia-ISO(N)[17].
Description[17]; Illustration[17].
Sangkhachand 1150: 'Climbing' [4].

var. **subsessilis** (Craib)De Wit [17]

B. subsessilis Craib [17].
Climbing or not[4]; Shrub or Tree[17]; Perennial[17].
Indo-China: Cambodia(N)[17]; Thailand(N)[17]. Asia: Malaysia-ISO(N)[17].
Description[16,17]; Illustration[16,17].
Miscellaneous[16].
Smitinand 2891: 'Scandent' [4].

var. **velutina** (Benth.)K. & S.Larsen [17]

B. velutina Bak. [17]; *Phanera velutina* Benth. [17].
Not climbing[17]; Shrub or Tree[17]; Perennial[17].
Indo-China: Thailand(N)[17]. Asia: Burma(N)[17]; Malaysia-ISO(N)[17].
Description[17]; Illustration[17].

B. prainiana Craib [17]

B. polycarpa var. *kurzii* Prain [17].
Not climbing[17]; Shrub[17]; Perennial[17].
Indo-China: Thailand(N)[17]. Asia: Burma(N)[17].
Description[17]; Illustration[17].

B. pulla Craib [16]

Climbing[16]; Shrub[16]; Perennial[16].
Indo-China: Cambodia(N)[16]; Thailand(N)[16].
Description[16,17]; Illustration[16,17].
Fibre[16].

B. purpurea L. [17]

B. castrata Blanco [16]; *B. coromandeliana* DC. [16]; *B. triandra* Roxb. [16]; *Phanera purpurea* (L.)Benth. [17].
Not climbing[17]; Shrub or Tree[17]; Perennial[17].
Indo-China: Laos(I)[16]; Thailand(I)[17]; Vietnam(I)[16].
Description[16,17]; Illustration[16,17].
Cultivated throughout the tropics. Originally palaeotropic [17].
The commonest cultivated form is *B.* × *blakeana* (*B.purpurea* × *B.variegata*) which is almost sterile and produces very few pods.

B. pyrrhoclada Drake [16]

Climbing[16]; Shrub[16]; Perennial[16].
Indo-China: Vietnam(N)[16]. Asia: China(N)[16].
Description[16]; Illustration[16].

B. racemosa Lam. [17]

B. parviflora Vahl [16]; *Piliostigma racemosa* (Lam.)Benth. [17].
Not climbing[17]; Tree[17]; Perennial[17].
Indo-China: Cambodia(N)[16]; Thailand(N)[17]; Vietnam(N)[16]. Asia: Burma(N)[17]; China (N)[17]; India(N)[17].
Description[16,17]; Illustration[16,17].

B. ridleyi Prain [17]

Phanera dasycarpa var. *ridleyi* (Prain)De Wit [17].
Indo-China: Thailand(N)[17]. Asia: Malaysia-ISO(N)[17].
Climbing[17]; Shrub[17]; Perennial[17].
Description[17]; Illustration[17].

B. rubro-villosa K. & S.Larsen [16]

B. mirabilis Gagnep. [16].
Climbing[16]; Shrub[16]; Perennial[16].
Indo-China: Laos(N)[16]; Vietnam(N)[16].
Description[16]; Illustration[16].
Medicine[16].

B. saccocalyx Pierre [17]

Not climbing[17]; Shrub or Tree[17]; Perennial[17].
Indo-China: Laos(N)[17]; Thailand(N)[17].
Description[16,17]; Illustration[16,17].

B. saigonensis Gagnep. [16]

Climbing[16]; Shrub[16]; Perennial[16].
Indo-China: Cambodia(N)[16]; Laos(N)[16]; Thailand(N)[16]; Vietnam(N)[16].
Description[16,17]; Illustration[16,17].
Domestic[16]; Fibre[16].
Thai material very poor [17].

var. **saigonensis** [16]

Climbing[16]; Endangered[16]; Shrub[16]; Perennial[16].
Indo-China: Vietnam(N)[16].
Description[16]; Illustration[16].

var. **gagnepainiana** K. & S.Larsen [16]

B. comosa Gagnep. [16].
Climbing[16]; Shrub[16]; Perennial[16].
Indo-China: Cambodia(N)[16]; Laos(N)[16]; Thailand(N)[16].
Description[16,17]; Illustration[17].
Domestic[16]; Fibre[16].

var. **poilanei** K. & S.Larsen [16]

Climbing[16]; Shrub[16]; Perennial[16].
Indo-China: Vietnam(N)[16].
Description[16].

B. scandens L. [16]

B. anguina Roxb. [16]; *B. debilis* Hassk. [16]; *Lasiobema anguinum* (Roxb.)Miq. [16]; *Lasiobema scandens* (L.)De Wit [16]; *Phanera bifoliata* Miq. [16]; *Phanera debilis* (Hassk.)Miq. [16]; *Phanera scandens* (L.)Raf. [16].
Climbing[16]; Shrub[16]; Perennial[16].
Indo-China: Cambodia(N)[16]; Laos(N)[16]; Thailand(N)[16]; Vietnam(N)[16]. Asia: China(N)[16]; India(N)[16]; Indonesia-ISO(N)[16].
Description[16,17]; Illustration[16,17].
Fibre[16]; Medicine[4].
Lakshnakara 255: 'Stem used as a relief for drunkness' [4].
Western area of the distribution: var. *scandens*. Oriental area of distribution: var. *horsfieldii* [16].

var. **horsfieldii** (Miq.)K. & S.Larsen [16]

B. anguina var. *horsfieldii* (Miq.)Prain [16]; *B. horsfieldii* (Miq.)Macbride [16]; *B. subrhombicarpa* Merr. [16]; *Lasiobema horsfieldii* Miq. [16]; *Lasiobema scandens* var. *horsfieldii* (Prain)De Wit [16].
Climbing[16]; Shrub[16]; Perennial[16].
Indo-China: Cambodia(N)[16]; Laos(N)[16]; Thailand(N)[16]; Vietnam(N)[16]. Asia: China(N)[16]; India(N)[16]; Indonesia-ISO(N)[16].
Description[16,17]; Illustration[16,17].
Fibre[16]; Medicine[4].
Monkey's Ladder[4].
Khoon Winit 496 'Monkey's Ladder' [4].

B. similis Craib [16]

Phanera similis (Craib)De Wit [16].
Climbing[16]; Shrub[16]; Perennial[16].
Indo-China: Laos(N)[16]; Thailand(N)[16]. Asia: Burma(N)[16].
Description[16,17]; Illustration[17].

B. strychnifolia Craib [17]

B. strychnifolia var. *pubescens* Craib [17].
Climbing or not[17]; Shrub[17]; Perennial[17].
Indo-China: Thailand(N)[17].
Description[17]; Illustration[17].
Fibre[4]; Food or drink[149].
Winit 1498: 'Bark used for cordage' [4].
Endemic to Thailand [17].

B. tomentosa L. [17]

B. pubescens DC. [16].
Not climbing[17]; Shrub[17]; Perennial[17].
Indo-China: Thailand(I)[17]; Vietnam(I)[17].
Description[16,17]; Illustration[16,17].
Cultivated throughout the tropics [17]. Found wild in Africa and India.

B. touranensis Gagnep. [16]

B. genuflexa Craib [16]; *B. glauca* sensu auct. [16]; *B. henryi* Craib [16]; *B. rocheri* A.Leveille [16].
Climbing[16]; Shrub[16]; Perennial[16].
Indo-China: Laos(N)[16]; Vietnam(N)[16]. Asia: Burma(N)[16]; China(N)[16].
Description[16]; Illustration[16].

B. tubicalyx Craib [17]

Lasiobema tubicalyx (Craib)De Wit [17].
Climbing[17]; Endangered[17]; Shrub[17]; Perennial[17].
Indo-China: Thailand(N)[17].
Description[17]; Illustration[17].
Endemic to Thailand [17].

B. variegata L. [17]

B. candida Aiton [16]; *Phanera variegata* (L.)Benth. [17].
Not climbing[17]; Shrub or Tree[17]; Perennial[17].
Indo-China: Laos(N)[17]; Thailand(N)[17]; Vietnam(N)[17]. Asia: Burma(U)[17]; China(U)[17]; India(U)[17].
Description[16,17]; Illustration[16,17].
Food or Drink[16]; Medicine[16].
Mountain Ebony[17]; St. Thomas Tree[17].
Cultivated in the tropics [16].
See note under *B. purpurea*.

B. viridescens Desv. [17]
Not climbing[16]; Shrub[16]; Perennial[16].
Indo-China: Cambodia(N)[16]; Laos(N)[16]; Thailand(N)[16]; Vietnam(N)[16]. Asia: Burma(N)[16]; China(N)[16]; Indonesia-ISO(N)[16].
Description[16,17]; Illustration[16,17].
Food or Drink[16]; Medicine[16].
A variable species [16].

var. **viridescens** [17]
B. baviensis Drake [16]; *B. polycarpa* Benth. [16]; *B. timorana* Decne.; *B. viridescens* var. *baviensis* (Drake)De Wit [16].
Not climbing[17]; Shrub[17]; Perennial[17].
Indo-China: Cambodia(N)[17]; Laos(N)[17]; Thailand(N)[17]; Vietnam(N)[17]. Asia: Burma(N)[17]; China(N)[17]; Indonesia-ISO(N)[17].
Description[17]; Illustration[17].

var. **hirsuta** K. & S.Larsen [17]
Not climbing[17]; Endangered[17]; Shrub[17]; Perennial[17].
Indo-China: Thailand[17].
Description[17].
Endemic to Thailand [17].

B. wallichii J.F.Macbr. [16]
B. melanophylla Merr. [16]; *B. macrostachya* (Benth.)Baker [16]; *Phanera macrostachya* Benth. [16].
Climbing[16]; Shrub[16]; Perennial[16].
Indo-China: Vietnam(N)[16]. Asia: Burma(N)[16]; India(N)[16].
Description[16]; Illustration[16].

B. winitii Craib [17]
Lysiphyllum winitii (Craib)De Wit [17].
Climbing[17]; Endangered[17]; Shrub[17]; Perennial[17].
Indo-China: Thailand(N)[17].
Description[17]; Illustration[17].
Endemic to Thailand [17].

B. yunnanensis Franchet [17]
B. altefissa A.Leveille [17]; *B. collettii* Prain [17]; *B. diptera* Coll. & Hemsley [17].
Climbing[17]; Shrub[17]; Perennial[17].
Indo-China: Thailand(N)[17]. Asia: Burma(N)[17]; China(N)[17].
Description[17]; Illustration[17].

DETARIEAE

CRUDIA Schreber

Trees, usually of forests and often near rivers. About 50 species, scattered through the tropics, but the majority in southeast Asia and Malesia.

C. caudata Prain [17]
Not climbing[17]; Tree[17]; Perennial[17].
Indo-China: Thailand(N) [17]. Asia: Malaysia-ISO(N) [17].
Description[17].

C. chrysantha (Pierre)Schumann [17]
C. chrysantha var. *harmandii* (Pierre)Gagnep. [17]; *Apalatoa chrysantha* Pierre [17]; *Apalatoa chrysantha* var. *harmandii* Pierre [17].
Not climbing[17]; Tree[17]; Perennial[17].
Indo-China: Cambodia(N) [17]; Laos(N) [17]; Thailand(N) [17]; Vietnam(N) [17].
Description[16,17]; Illustration[16,17].
Wood[17].

C. evansii Ridl. [17]
Not climbing[17]; Shrub or Tree[17]; Perennial[17].
Indo-China: Thailand(N) [17]. Asia: Malaysia-ISO(N) [17].
Description[17]; Illustration[17].
Closely related to *C. bantamensis* (Hassk.)Benth. from Java [17].

C. gracilis Prain [17]
C. brevipes Ridl. [17].
Not climbing[17]; Shrub[17]; Perennial[17].
Indo-China: Thailand(N) [17]. Asia: Malaysia-ISO(N) [17].
Description[17].

C. lanceolata Ridl. [17]
C. curtisii sensu auct. [17].
Indo-China: Thailand[17]. Asia: Malaysia-ISO(N) [17].
Not climbing[17]; Tree[17]; Perennial[17].
Description[17]; Illustration[17].

C. speciosa Prain [17]
Not climbing[17]; Tree[17]; Perennial[17].
Indo-China: Thailand(N) [17].
Description[17].
Endemic to Thailand [17].

CYNOMETRA L.

A genus of perhaps 70 species, scattered through the tropics, mostly in forest; most diverse in Africa. A few yield timber, and the young fruits of *C. cauliflora* are used as a vegetable.

C. cauliflora L. [17]
Not climbing[17]; Tree[17]; Perennial[17].
Indo-China: Thailand(I) [17]. Asia: Indonesia-ISO(N) [17].
Description[17].
Food or Drink[17].
Cultivated but not common in Thailand. Only known as cultivated on Asiatic mainland [17].

C. craibii Gagnep. [16]
Not climbing[17]; Tree[17]; Perennial[17].
Indo-China: Laos(N) [17]; Thailand(N) [17].
Description[16,17]; Illustration[16,17].
Wood[17].

C. dongnaiensis Pierre [16]
Not climbing[16]; Tree[16]; Perennial[16].
Indo-China: Cambodia(N) [16]; Vietnam(N) [16]. Asia: Philippines(N) [21].
Description[16]; Illustration[16].
Wood[16].

C. glomerulata Gagnep. [16]
Not climbing[16]; Tree[16]; Perennial[16].
Indo-China: Laos(N) [16]; Vietnam(N) [16].
Description[16]; Illustration[16].

C. iripa Kostel. [17]
C. bijuga sensu auctt. [17]; *C. bijuga* var. *mimosoides* (Bak.)Merr. [17]; *C. mimosoides* (Bak.)Prain [17]; *C. ramiflora L. var. mimosoides* Baker [16].
Not climbing[17]; Shrub or Tree[17]; Perennial[17].
Indo-China: Thailand(N) [17]. Asia: Burma(N) [17]; India(N) [17]; Indonesia-ISO(N) [17]; Malaysia-ISO(N) [17]; Philippines(N) [17]; Sri Lanka(N) [17]. Australasia: Australia(N) [17].
Description[17].

C. malaccensis Van Meeuwen [17]

C. inaequalifolia Bak. [17].
Not climbing[17]; Tree[17]; Perennial[17].
Indo-China: Thailand(N) [17]. Asia: India(N) [17]; Malaysia-ISO(N) [17].
Description[17]; Illustration[17].
Wood[17].

INTSIA Thouars

Three species, all trees, in Asia and on the coasts of the Indian and Pacific Oceans. Some yield high-quality timber.

I. bijuga (Colebr.)Kuntze [17]

I. amboinensis DC. [16]; *I. cambodiensis* (Hance)Pierre [17]; *I. madagascariensis* DC. [16]; *I. retusa* (Kurz)Kuntze [17]; *Afzelia bijuga* (Colebr.)A.Gray [17]; *Afzelia cambodiensis* Hance [17]; *Afzelia retusa* Kurz [17]; *Jonesia triandra* Roxb. [16]; *Macrolobium bijugum* Colebr. [16]; *Outea bijuga* (Colebr.)DC. [16]; *Pahudia hasskarliana* Miq. [17].
Not climbing[17]; Tree[17]; Perennial[17].
Indo-China: Cambodia(N) [17]; Thailand(N) [17]; Vietnam(N) [17]. Asia: Burma(N) [17]; India(N) [17]; Indonesia-ISO(N) [17]; Malaysia-ISO(N) [17]; Australasia; Australia(N) [17]; Indian Ocean: Madagascar(N) [17].
Description[16,17]; Illustration[16,17].

Intsia palembanica Miq. [17]

I. bakeri (Prain)Prain [17]; *I. plurijuga* Harms [17]; *Afzelia bakeri* Prain [17]; *Afzelia palembanica* (Miq.)Bak. [17].
Not climbing[17]; Tree[17]; Perennial[17].
Indo-China: Thailand(N) [17]. Asia: Burma(N) [17]; Indonesia-ISO(N) [17]; Malaysia-ISO(N) [17].
Description[17].
Wood[17].

LYSIDICE Hance

A single species. Southeast Asian in origin, but planted elsewhere for its decorative flowers. A small tree or shrub.

L. rhodostegia Hance [16]

Indo-China: Vietnam(N) [16]. Asia: China(N) [16]; Hong Kong(N) [16].
Not climbing[16]; Shrub[16]; Perennial[16].
Description[16]; Illustration[16].

MANILTOA R.Scheffer

A genus of 20 to 25 species of trees, occurring in SE Asia and Malesia, with most species in New Guinea.

M. polyandra (Roxb.)Harms [16]

Cynometra polyandra Roxb. [16].
Not climbing[16]; Tree[16]; Perennial[16].
Indo-China: Cambodia(N) [16]; Laos(N) [16]; Thailand(N) [16]. Asia: Bangladesh(N) [16]; Burma(N) [16]; India(N) [16]; Malaysia-ISO(N) [16].
Description[16]; Illustration[16].
Wood[16].

PHYLLOCARPUS Endl.

A genus of two tree species, occurring naturally in South America.

P. septentrionalis J.Donn. Smith [17]
Not climbing[17]; Tree[17]; Perennial[17].
Indo-China: Thailand(I) [17].
Description[17].
Environmental[17].

SARACA L.

A genus of about 10 species from the Indo-Malaysian region. All are trees.

S. declinata (Jack)Miq. [17]
S. biglandulosa Pierre [17]; *S. cauliflora* Bak. [17]; *S. macroptera* Miq. [17]; *S. macroptera* var. *paucijuga* Craib [17]; *S. macroptera* var. *siamensis* Craib [17]; *S. thorelii* Gagnep. [17]; *S. triandra* Bak. [17]; *S. zollingeriana* Miq. [16]; *Jonesia declinata* Jack [16].
Not climbing[17]; Tree[17]; Perennial[17].
Indo-China: Cambodia(N) [16]; Laos(N) [17]; Thailand(N) [17]; Vietnam(N) [17]. Asia: Burma(N) [17]; Indonesia-ISO(N) [17]; Malaysia-ISO(N) [17].
Description[16,17]; Distribution Map[23]; Illustration[16,17].
Environmental[17]; Food or Drink[17]; Medicine[17]; Wood[17].

S. dives Pierre [16]
Not climbing[16]; Shrub[16]; Perennial[16].
Indo-China: Laos(N) [16]; Vietnam(N) [16].
Description[16]; Distribution Map[23]; Illustration[16].
Environmental[16]; Food or Drink[16].
Very closely related to *S.thaipingensis* Prain [16].

S. indica L. [16]
S. asoca sensu auctt. [16]; *S. bijuga* Prain [16]; *S. harmandiana* Pierre [16]; *S. indica* var. *bijuga* (Prain)Gagnep. [16]; *S. indica* var. *zollingeriana* (Miq.)Gagnep. [16]; *S. minor* (Zoll. & Mor.)Miq. [16]; *S. pierreana* Craib [16]; *Jonesia asoca* sensu auctt. [16]; *Jonesia minor* Zoll. & Mor. [16].
Not climbing[16]; Tree[16]; Perennial[16].
Indo-China: Laos(N) [16]; Thailand(N) [16]; Vietnam(N) [16]. Asia: Indonesia-ISO(N) [16]; Malaysia-ISO(N) [16].
Description[16,17]; Distribution Map[23]; Illustration[16,17].
S. asoca (Roxb.)Wilde is limited to India [16].

S. schmidiana J.E.Vidal [16]
Not climbing[16]; Tree[16]; Perennial[16].
Indo-China: Vietnam(N) [16].
Description[16]; Illustration[16].
Species known only from the type [16].

S. thaipingensis Prain [17]
S. declinata sensu auct. [17].
Not climbing[17]; Tree[17]; Perennial[17].
Indo-China: Thailand(N) [17]; Vietnam(N) [23]. Asia: Burma(N) [17]; Indonesia-ISO(N) [17]; Malaysia-ISO(N) [17].
Description[17]; Distribution Map[23]; Illustration[17].
Environmental[17].

SINDORA Miq.

Trees; about 20 species in southeast Asia and Malesia, but one in Africa. Some yield timber.

S. coriacea (Bak.)Prain [17]

Afzelia coriacea Bak. [17].
Not climbing[17]; Tree[17]; Perennial[17].
Indo-China: Thailand(N) [17]. Asia: Indonesia-ISO(N) [17]; Malaysia-ISO(N) [17].
Description[17]; Illustration[17].
Wood[17].

S. echinocalyx Prain [17]

S. fusca Craib [17].
Not climbing[17]; Tree[17]; Perennial[17].
Indo-China: Thailand(N) [17]. Asia: Malaysia-ISO(N) [17].
Description[17].
Wood[17].

S. laotica Gagnep. [16]

S. koutumensis Gagnep. [16].
Not climbing[16]; Tree[16]; Perennial[16].
Indo-China: Laos(N) [16]; Vietnam(N) [16].
Description[16]; Illustration[16].

S. siamensis Miq. [17]

Not climbing[17]; Tree[17]; Perennial[17].
Indo-China: Cambodia(N) [17]; Laos(N) [17]; Thailand(N) [17]; Vietnam(N) [17]. Asia: Malaysia-ISO(N) [17].
Description[16,17]; Illustration[16,17].
Chemical Products[22]; Domestic[22]; Food or Drink[16]; Wood[17].
Dominant species in dry deciduous dipterocarp forests [17].

var. **siamensis** [17]

Galedupa cochinchinensis (Baillon)Prain [16]; *Galedupa siamensis* (Teijsm.)Prain [16]; *S. cochinchinensis* Baillon [17]; *S. wallichii* var. *siamensis* (Teijsm.)Bak. [17].
Not climbing[17]; Tree[17]; Perennial[17].
Indo-China: Cambodia(N) [17]; Laos(N) [17]; Thailand(N) [17]; Vietnam(N) [17]. Asia: Malaysia-ISO(N) [17].
Description[16,17]; Illustration[16,17].
Food or Drink[16]; Wood[17].
Dominant in dry deciduous dipterocarp forests [17].

var. **maritima** (Pierre)K. & S. Larsen [17]

S. cochinchinensis var. *maritima* (Pierre)De Wit [16]; *S. maritima* Pierre [16].
Not climbing[16]; Tree[16]; Perennial[16].
Indo-China: Cambodia(N) [16]; Laos(N) [16]; Thailand(N) [17]; Vietnam(N) [16].
Description[16,17]; Illustration[16,17].
Environmental[4]; Food or Drink[16]; Wood[16].
Collins s.n: 'ornamental' [4].

S. tonkinensis K. & S.Larsen [16]

Not climbing[16]; Tree[16]; Perennial[16].
Indo-China: Cambodia(N) [16]; Vietnam(N) [16].
Description[16]; Illustration[16].
Domestic[16]; Wood[16].
Presence of this species in Cambodia is doubtful; only one specimen, perhaps wrongly labelled [16].

MIMOSOIDEAE

ACACIEAE

ACACIA Miller

A very large pantropical genus with over 1000 species, mainly in dry country. In Indo-China, however, most species are spiny lianas. The spineless Australian species, with phyllodes instead of leaves, are widely introduced in tropical and subtropical regions, but few have been recorded from Indo-China, probably because of under-collecting of cultivated plants.

A. andamanica I.Nielsen [1]

A. pseudo-intsia sensu auct. [1]; *A. pseudo-intsia* Miq. var. *ambigua* Prain [1].
Climbing [1]; Shrub [1]; Perennial [1].
Indo-China: Thailand(N) [1]. Asia: India (Andaman Islands)(N) [1].
Description [1]; Illustration [1].

A. caesia (L.)Willd. [2]

A. columnaris Craib [2]; *A. intsia* sensu auct. [2]; *A. intsia* var. *caesia* (L.)Bak. [2]; *Mimosa caesia* L. [2].
Climbing or not [2]; Shrub [2]; Perennial [2].
Indo-China: Cambodia(N) [2]; Laos(N) [2]; Thailand(N) [1]; Vietnam(N) [2]. Asia: Burma(N) [1]; India(N) [1].
Description [1,2]; Illustration [1,2].
Domestic [2]; Medicine [2].
Var. *caesia* occurs in India and Sri Lanka and does not reach Thailand [1].

var. **subnuda** (Craib)I.Nielsen [1]

A. oxyphylla Benth. [1]; *A. oxyphylla* var. *subnuda* Craib [1].
Climbing or not [1]; Shrub [1]; Perennial [1].
Indo-China: Cambodia(N) [2]; Laos(N) [2]; Thailand(N) [1]; Vietnam(N) [2]. Asia: Burma(N) [1]; China(N) [147]; India(N) [1].
Description [1,2]; Illustration [1,2].
Domestic [2]; Medicine [2].
Var. *caesia* occurs in India and Sri Lanka and does not reach Thailand [1].

A. catechu (L.f.)Willd. [1]

A. sundra (Roxb.)Bedd. [1]; *Mimosa catechu* L.f. [1].
Not climbing [4]; Tree [4]; Perennial [4].
Indo-China: Thailand(U) [4]. Asia: Burma(N) [1]; India(N) [1]; Sri Lanka(N) [1].
An introduction; recorded only once, from Chiang Mai, Thailand [4]; *Kerr* 1230 [4].

A. comosa Gagnep. [2]

Climbing [1]; Shrub [1]; Perennial [1].
Indo-China: Cambodia(N) [2]; Laos(N) [1]; Thailand(N) [1]; Vietnam(N) [1].
Description [1,2]; Illustration [1,2].
The only species in Indo-China with alternate leaflets [2].

A. concinna (Willd.)DC. [2]

A. concinna var. *rugata* (Benth.)Bak. [2]; *A. pennata* sensu auct. [5]; *A. poilanei* Gagnep. [6]; *A. pseudo-intsia* sensu auctt. [6]; *A. quisumbingii* Merr. [6]; *A. rugata* (Lam.)Merr. [6]; *Mimosa concinna* Willd. [2].
Not climbing [1]; Shrub [1]; Perennial [1].
Indo-China: Cambodia(N) [2]; Laos(N) [2]; Thailand(N) [1]; Vietnam(N) [2]. Asia: India(N) [1]; Indonesia-ISO(N) [6]; Malaysia-ISO(N) [6]; Philippines(N) [6].
Description [1,2]; Illustration [1,2].
Domestic [1]; Food or Drink [2].
Rarely a low tree [1].

A. confusa Merr. [2,4]
Not climbing [4]; Shrub or Tree[4]; Perennial [4].
Indo-China: Vietnam(I) [4]. Asia: Philippines(N) [2]; Taiwan(N) [2].

A. craibii I.Nielsen [1]
Nimiria siamensis Craib [1].
Not climbing [1]; Shrub [1]; Perennial [1].
Indo-China: Thailand(N) [1].
Description [1]; Illustration [1].
Endemic to Thailand [1].

A. donnaiensis Gagnep. [2]
Climbing [2]; Shrub [2]; Perennial [2].
Indo-China: Cambodia(N) [2]; Vietnam(N) [2]. Asia: Indonesia-ISO(N) [6]; Malaysia-ISO(N) [2],Borneo [2].
Description [2]; Distribution Map [6]; Illustration [2].
Environmental [2]; Wood [2].

A. farnesiana (L.)Willd. [2]
Mimosa farnesiana L. [2]; *Vachellia farnesiana* (L.)Wight & Arn. [2].
Not climbing [1]; Shrub or Tree[1]; Perennial [1].
Indo-China: Cambodia(I) [2]; Laos(I) [2]; Thailand(I) [1]; Vietnam(I) [2].
Description [1,2]; Illustration [1,2].
Domestic [1]; Environmental [1]; Medicine [2].
Sponge Tree [1].
Origin uncertain; probably South American. A perfume is extracted from the flowers [2].

A. harmandiana (Pierre)Gagnep. [2]
A. siamensis Craib [2]; *Delaportea armata* Gagnep. [2]; *Pithecolobium mekongense* Pierre [2].
Not climbing [1]; Tree [1]; Perennial [1].
Indo-China: Cambodia(N) [2]; Laos(N) [1]; Thailand(N) [1]; Vietnam(N) [1].
Description [1,2]; Illustration [1,2].
Miscellaneous [1]; Wood [1].

A. leucophloea (Roxb.)Willd. [2]
Delaportea ferox Gagnep. [2]; *Delaportea microphylla* Gagnep. [2]; *Mimosa leucophloea* Roxb. [2].
Not climbing [1]; Tree [1]; Perennial [1].
Indo-China: Thailand(N) [1]; Vietnam(N) [2]. Asia: Burma(N) [1]; India(N) [1]; Indonesia-ISO(N) [1].
Description [1,2]; Illustration [1,2].
Domestic [1]; Wood [1].

A. meeboldii Craib [1]
Climbing [1]; Shrub [1]; Perennial [1].
Indo-China: Thailand(N) [1]. Asia: Burma(N) [1].
Description [1]; Illustration [1]
Largest leaflets known in *Acacia* in Asia [1].

A. megaladena Desv. [2]
Climbing [2]; Shrub [2]; Perennial [2].
Indo-China: Cambodia(N) [1]; Laos(N) [2]; Thailand(N) [2]; Vietnam(N) [2]. Asia: Burma(N) [2]; China(N) [2]; India(N) [2]; Indonesia-ISO(N) [2].
Description [1,2]; Distribution Map [6]; Illustration [1,2].
Domestic [2].

var. **megaladena** [2]
A. pennata sensu auctt. [2]; *A. pennata* var. *arrophula* (D.Don)Bak. [2]
Climbing [2]; Shrub [2]; Perennial [2].
Indo-China: Laos(N) [2]; Thailand(N) [2]; Vietnam(N) [2]. Asia: Burma(N) [2]; China(N) [2]; India(N) [2]; Indonesia-ISO(N) [2].
Description [1,2]; Distribution Map [6]; Illustration [1,2].

var. **garrettii** I.Nielsen [1]
Climbing [1]; Shrub [1]; Perennial [1].
Indo-China: Thailand(N) [1]. Asia: China(N) [1].
Description [1]; Distribution Map [6]; Illustration [1].

var. **indo-chinensis** I.Nielsen [2]
A. pennata var. *arrophula* sensu auctt. [2].
Climbing [2]; Shrub [2]; Perennial [2].
Indo-China: Cambodia(N) [2]; Laos(N) [2]; Thailand(N) [2]; Vietnam(N) [2]. Asia: Indonesia-ISO(N) [6].
Description [1,2]; Distribution Map [6]; Illustration [1,2].
Domestic [2]; Toxins [2].
Craib [Fl.Siam.Enum.1(3):551(1928)] distinguished this as a variety but did not give it a name [2].

A. pennata (L.)Willd. [2]
Mimosa pennata L. [2].
Climbing [1]; Shrub [1]; Perennial [1].
Indo-China: Cambodia(N) [1]; Laos(N) [1]; Thailand(N) [1]; Vietnam(N) [1]. Asia: Burma(N) [1]; China(N) [1]; India(N) [1]; Indonesia-ISO(N) [6]; Malaysia-ISO(N) [6]; Sri Lanka(N) [1].
Description [1,2]; Distribution Map [6]; Illustration [1,2].
Environmental [1]; Food or Drink [1]; Medicine [2].
Cultivated in Thailand, Laos and Cambodia [1].

subsp. **pennata** [2]
A. pennata var. *canescens* Kurz [2].
Indo-China: Thailand(N) [2]. Asia: Burma(N) [2]; India(N) [2]; Sri Lanka(N) [2].
Climbing [2]; Shrub [2]; Perennial [2].
Description [1,2]; Distribution Map [6]; Illustration [1,2].
The main distribution of this subspecies is in Southern India and Sri Lanka [1]

subsp. **hainanensis** (Hayata)I.Nielsen [2]
A. hainanensis Hayata [2]; *A. macrocephala* Lace [2].
Climbing [2]; Shrub [2]; Perennial [2].
Indo-China: Vietnam(N) [2]. Asia: Burma(N) [2]; China(N) [2].
Description [2]; Illustration [2].

subsp. **insuavis** (Lace)I.Nielsen [1]
A. insuavis Lace [1].
Climbing [1]; Shrub [1]; Perennial.
Description [1,2]; Illustration [1,2].
Environmental [1]; Food or Drink [1]; Medicine [2].
Cultivated in Thailand, Laos and Cambodia [1].

subsp. **kerrii** I.Nielsen [2]
Climbing [2]; Shrub [2]; Perennial [2].
Indo-China: Cambodia(N) [1]; Laos(N) [1]; Thailand(N) [1]; Vietnam(N) [1] Asia: Burma(N) [1]; India(N) [1]; Indonesia-ISO(N) [1]; Malaysia-ISO(N) [1].
Description [1,2]; Distribution Map [6]; Illustration [1,2].
The main distribution of this subspecies is in Thailand and Indo-China [1].

A. pluricapitata Benth. [2]
A. pennata var. *pluricapitata* (Steud.)Bak. [2].
Climbing [2] ; Shrub [2] ; Perennial [2]
Indo-China: Thailand(N) [1]; Vietnam(N) [1] Asia: Indonesia-ISO(N) [1]; Malaysia-ISO(N) [1].
Description [1,2,6]; Distribution Map [6]; Illustration [1,2,6].
Medicine [6].

A. pruinescens Kurz [2]
Climbing [2]; Shrub [2]; Perennial [2].
Indo-China: Vietnam(N) [2]. Asia: Burma(N) [2]; China(N) [2].
Description [2]; Illustration [2].

A. pseudo-intsia Miq. [1]
A. macrocephala var. siamensis Craib [1].
Climbing or not [1]; Shrub [1]; Perennial [1].
Indo-China: Thailand(N) [1]. Asia: Indonesia-ISO(N) [1]; Malaysia-ISO(N) [1].
Description [1]; Distribution Map [6]; Illustration [1].

A. thailandica I.Nielsen [2]
Climbing or not [1]; Shrub [1]; Perennial [1].
Indo-China: Cambodia(N) [1]; Thailand(N) [1].
Description [1,2]; Illustration [1,2].

A. tomentosa Willd. [2]
A. chrysocoma Miq. [2]; *A. tomentosa* var. *chrysocoma* (Miq.)Back. [2]; *Mimosa tomentosa* (Willd.)Roxb. [2].
Not climbing [1]; Shrub or Tree[1]; Perennial [1].
Indo-China: Thailand(N) [1]; Vietnam(N) [2]. Asia: Burma(N) [1]; India(N) [1]; Indonesia-ISO(N) [1].
Description [1,2]; Illustration [1,2].
Wood [1].

A. tonkinensis I.Nielsen [2]
Climbing [2]; Shrub [2]; Perennial [2].
Indo-China: Vietnam(N) [2].
Description [2]; Illustration [2].
This species is known only from northern Vietnam [2].

A. torta (Roxb.)Craib [1]
Mimosa torta Roxb. [1].
Climbing or not [1]; Shrub [1]; Perennial [1].
Indo-China: Thailand(N) [1]. Asia: India(N) [1].
Description [1]; Illustration [1].

A. vietnamensis I.Nielsen [2]
Climbing [2]; Shrub [2]; Perennial [2].
Indo-China: Laos(N) [2]; Vietnam(N) [2].
Description [2]; Illustration [2].

A. sp. I.Nielsen [1]
Indo-China: Thailand(N) [1].
Description [1].
Characters of the pod and seeds suggest that this is closely related to *A. megaladena* [1]. Flowering material will show if this entity deserves specific rank [1].
Sorensen, Larsen & Hansen 2205(c) [1].

INGEAE

ALBIZIA Duraz.

A large (c. 150 spp.) genus of woody tropical plants, found throughout the tropics, mainly in wetter habitats but by no means confined to them. A few species extend into the warm temperate zone. Some yield useful timber; others are ornamental.

A. attopeuensis (Pierre)I.Nielsen [2]
Pithecolobium attopeuense Pierre [2]; *Pithecellobium corymbosum* sensu Gapnep, p.p. [2]; *Serialbizzia attopeuense* (Pierre)Kosterm. [2].
Not climbing[2]; Tree[2]; Perennial[2].
Indochina: Laos(N) [2]; Thailand(N) [2]; Vietnam(N) [2]. Asia: China(N) [2].
Description [1,2]; Illustration [1,2].
Timber; Food or drink.

A. chinensis (Osbeck)Merr. [2]

A. marginata (Lam.)Merr. [2]; *A. stipulata* (DC.)Boivin [2]; *Mimosa chinensis* Osbeck [2]; *Mimosa marginata* Lam. [2].
Not climbing[2]; Tree[2]; Perennial[2].
Indo-China: Cambodia(N) [2]; Laos(N) [2]; Thailand(N) [1]; Vietnam(N) [2]. Asia: Burma(N) [2]; China(N) [2]; India(N) [2]; Indonesia-ISO(N) [6]; Malaysia-ISO(N) [6]; Sri Lanka(N) [2].
Description[1,2]; Illustration[1,2].
Environmental[1].
Cultivated [1].

A. corniculata (Lour.)Druce [2]

A. millettii Benth. [2]; *A. millettii* var. *arfeuilleana* Gagnep. [2]; *A. millettii* var. *siamensis* Craib [2]; *A. nigricans* Gagnep. [2]; *A. scandens* Merr. [2]; *Caesalpinia lebbekioides* DC. [2]; *Mimosa corniculata* Lour. [2].
Climbing or not[1]; Shrub or tree[1]; Perennial[1].
Indo-China: Cambodia(N) [2]; Laos(N) [2]; Thailand(N) [1]; Vietnam(N) [2]. Asia: China(N) [1]; Malaysia-ISO(N) [1]; Borneo[1]; Philippines(N) [1].
Description[1,2]; Distribution Map[6,9]; Illustration[1,2].
Domestic[2].

A. crassiramea Lace [2]

A. laotica Gagnep. [2]; *A. saponaria* sensu auct. [2].
Not climbing[2]; Tree[2]; Perennial[2].
Indo-China: Laos(N) [2]; Thailand(N) [2]; Vietnam(N) [2]. Asia: Burma(N) [2]; China(N) [1].
Description[1,2]; Distribution Map[9]; Illustration[1,2].

A. garrettii I.Nielsen [1]

Not climbing[1]; Tree[1]; Perennial[1].
Indo-China: Thailand(N) [1]. Asia: Burma(N) [1]; China(N) [1].
Description[1]; Distribution Map[9]; Illustration[1].

A. kalkora Prain [2]

A. esquirolli Leveille [2]; *A. henryi* Ricker [2]; *A. simeonis* Harms [2].
Not climbing[2]; Shrub or tree[2]; Perennial[2].
Indo-China: Vietnam(N) [2]. Asia: China(N) [2]; Japan(N) [2].
Description[2]; Illustration[2].

A. lebbeck (L.)Benth. [2]

A. latifolia Boivin [2]; *Acacia lebbek* (L.)Willd. [2]; *Acacia speciosa* (Jacq.)Willd. [2]; *Mimosa lebbek* L. [2]; *Mimosa sirissa* Roxb. [2]; *Mimosa speciosa* Jacq. [2].
Not climbing[2]; Tree[2]; Perennial[2].
Indo-China: Cambodia(I) [2]; Thailand(I) [1]; Vietnam(I) [2].
Description[1,2]; Illustration[1,2].
Environmental[2]; Wood[2].
Introduced and cultivated all over the tropics.
Probably native to tropical mainland Asia [6].

A. lebbekoides (DC.)Benth. [2]

Acacia lebbekoides DC. [2].
Indo-China: Cambodia(N) [1]; Laos(N) [1]; Thailand(N) [1]; Vietnam(N) [1]. Asia: Indonesia-ISO(N) [6]; Malaysia-ISO(N) [1].
Not climbing[2]; Tree[2]; Perennial[2].
Description[1,2]; Distribution Map[6]; Illustration[1,2].
Food or Drink[4]; Medicine[4]; Wood[1].

A. lucidior (Steud.)I.Nielsen [2]

A. bracteata Dunn [2]; *A. gamblei* Prain [2]; *A. lucida* Benth. [2]; *A. meyeri* Ricker [2]; *Inga lucidior* Steud. [2].
Not climbing[1]; Tree[1]; Perennial[1].
Indo-China: Cambodia(N) [2]; Laos(N) [2]; Thailand(N) [1]; Vietnam(N) [2]. Asia: Burma(N) [1]; China(N) [1]; India(N) [1].
Description[1,2]; Illustration[1,2].
Forage[1].
Also cultivated. Native to tropical mainland Asia, excluding Malaya [6].

A. myriophylla Benth. [2]

A. thorelii Pierre [2]; *A. vialeana* var. *thorelii* (Pierre)Pham Hoang Ho [2].
Climbing[1]; Shrub[2]; Perennial[2].
Indo-China: Cambodia(N) [2]; Laos(N) [2]; Thailand(N) [1]; Vietnam(N) [2]. Asia: Burma(N) [2]; India(N) [2].
Description[1,2]; Distribution Map[6,9]; Illustration[1,2].
Food or Drink[2]; Medicine[2].

A. odoratissima (L.f.)Benth. [2]

A. odoratissima var. *mollis* Bak. [2]; *Acacia odoratissima* (L.f.)Willd. [2]; *Mimosa odoratissima* L.f. [2].
Not climbing[2]; Tree[2]; Perennial[2].
Indo-China: Laos(N) [2]; Thailand(N) [1]; Vietnam(N) [2]. Asia: China(N) [2]; India(N) [2]; Sri Lanka(N) [2].
Description[1,2]; Illustration[1,2].
Medicine[2]; Wood[1].
Ceylon Rosewood[6].
Also cultivated [1].

A. poilanei I.Nielsen [2]

Not climbing[2]; Tree[2]; Perennial[2].
Indo-China: Vietnam(N) [2].
Description[2]; Distribution Map[9]; Illustration[2].
Endemic to South Vietnam [2].

A. procera (Roxb.)Benth. [2]

A. procera var. *elata* (Roxb.)Bak. [2]; *Acacia procera* (Roxb.)Willd. [2]; *Mimosa elata* Roxb. [2]; *Mimosa procera* Roxb. [2].
Not climbing[2]; Tree[2]; Perennial[2].
Indo-China: Cambodia(N) [2]; Laos(N) [2]; Thailand(N) [1]; Vietnam(N) [2]. Asia: Burma(N) [6]; China(N) [6]; India(N) [6]; Indonesia-ISO(N) [6]. Malaysia-ISO(N) [6]; Philippines(N) [6].
Description[1,2]; Illustration[1,2].
Wood[1].

A. retusa Benth. [1]

A. littoralis Teijsm.& Binn. [1].
Not climbing[1]; Tree[1]; Perennial[1].
Indo-China: Thailand(N) [1]. Asia: Indonesia-ISO(N) [6]; Malaysia-ISOa(N) [6]; Borneo[6]; Philippines(N) [6].
Description[1]; Distribution Map[6]; Illustration[1].
Of the two subspecies, only ssp. *retusa* occurs in Indo-China.
The other, ssp. *morobei* occurs in Papua New Guinea [6].

A. saponaria Miq. [4]

Not climbing[4]; Tree[4]; Perennial[4].
Indo-China: Vietnam(I) [4]. Asia: Malaysia-ISO(N) [4].
Bon 4881 [4]. Introduced and not naturalized.

A. vialeana Pierre [2]

Not climbing[2]; Tree[2]; Perennial[2].
Indo-China: Cambodia(N) [2]; Thailand(N) [2]; Vietnam(N) [2].
Description[1,2]; Distribution Map[9]; Illustration[1,2].

ARCHIDENDRON F. Muell.

A genus of about 40 species of forest trees, confined to tropical Asia and Malesia with its centre of diversity in New Guinea. Some are timber trees.

A. balansae (Oliver)I.Nielsen [2]

Albizia balansae (Oliver)Huang [10]; *Cylindrokelupha balansae* (Oliver)Kosterm.,p.p. [2]; *Ortholobium annamense* Gagnep.,p.p. [2]; *Ortholobium balansae* (Oliver)Gagnep. [2]; *Pithecellobium balansae* Oliver [2].
Not climbing[2]; Tree[2]; Perennial[2].
Indo-China: Vietnam(N) [2]. Asia: China(N) [2].
Description[2]; Distribution Map[8]; Illustration[2,11].
Only known from the centre and north of Vietnam [2].

A. bauchei (Gagnep.)I.Nielsen [2]

Abarema bauchei (Gagnep.)Kosterm. [2]; *Abarema clypearia* sensu auctt. [2]; *Pithecellobium bauchei* Gagnep. [2]; *Pithecellobium clypearia* var. *acuminatum* sensu auctt. [2].
Not climbing[2]; Shrub or tree[2]; Perennial[2].
Indo-China: Vietnam(N) [2].
Description[2]; Distribution Map[8]; Illustration[2].
Only known from Vietnam [2].

A. bubalinum (Jack)I.Nielsen [1]

Cylindrokelupha bubalina (Jack)Kosterm.,p.p. [1]; *Feuilleea bubalina* (Jack)Kuntze [1]; *Pithecellobium bubalinum* (Jack)Benth. [1].
Not climbing[1]; Tree[1]; Perennial[1].
Indo-China: Thailand(N) [1]. Asia: Indonesia-ISO(N) [1]; Malaysia-ISO(N) [1].
Description[1]; Distribution Map[10]; Illustration[1].
The testa of the seeds emits a strong smell of garlic [1].

A. chevalieri (Kosterm.)I.Nielsen [2]

Abarema robinsonii sensu Kosterm.,p.p.; *Abarema yunnanense* sensu Kosterm.,p.p.; *Albizia chevalieri* (Kosterm.)Huang [10]; *Cylindrokelupha chevalieri* Kosterm. [2]; *Cylindrokelupha platyphylla* sensu auct. [2]; *Cylindrokelupha robinsonii* sensu Kosterm. [2]; *Ortholobium chevalieri* Gagnep. [2]; *Ortholobium platyphyllum* Gagnep.,p.p. [2]; *Paralbizzia robinsonii* sensu Kosterm.,p.p. [2].
Not climbing[2]; Tree[2]; Perennial[2].
Indo-China: Vietnam(N) [2]. Asia: China(N) [2].
Description[2]; Distribution Map[8]; Illustration[2].
Most specimens cited under *Cylindrokelupha robinsonii* Kosterm. (1966) belong to this species [2].

A. clypearia (Jack)I.Nielsen [2]

Not climbing[1]; Shrub or tree[1]; Perennial[1].
Indo-China: Cambodia(N) [2]; Laos(N) [2]; Thailand(N) [1]; Vietnam(N) [2].
Description[1,2,10]; Distribution Map[10]; Illustration[1,2,10].
Two subspecies occur, *clypearia* & *subcoriaceum* [10].
Two varieties of subsp. *clypearia* occur in Indo-China, var. *clypearia* and var. *sessiliflorum*. Subsp. *subcoriaceum* occurs in India and Sri-Lanka [10].

subsp. **clypearia** [2]

Abarema angulata (Benth.)Kosterm. [2]; *Abarema clypearia* (Jack)Kosterm. [2]; *Abarema clypearia* var. *angulata* (Benth.)Kosterm. [2]; *Abarema cuneadena* (Kosterm.)Kosterm. [1]; *Abarema sessiliflora* (Merr.)Kosterm. [1]; *Abarema sessiliflora* subsp. *sessiliflora* (Merr.)I.Nielsen [1]; *Albizzia angulata* (Benth.)Kurz [2]; *Feuilleea clypearia* (Jack)Kuntze [2]; *Pithecellobium angulatum* Benth. [2]; *Pithecellobium clypearia* (Jack)Benth. [2]; *Pithecellobium clypearia* var. *acuminatum* Gagnep. [2]; *Pithecellobium cuneadenum* Kosterm. [1]; *Pithecellobium montanum* Benth. [2]; *Pithecellobium sessiliflorum* Merr. [1].
Not climbing[1,2]; Shrub or Tree[1,2]; Perennial[1,2].
Indo-China: Cambodia(N) [2]; Laos(N) [2]; Thailand(N) [1]; Vietnam(N) [2]. Asia: India(N) [2]; Malaysia-ISO(N) [1]; Philippines(N) [1].
Description[1,2]; Distribution Map[10]; Illustration[1,2,10].
Chemical Products[2]; Domestic[2]; Fibre[2]; Medicine[2].
Var. *sessiliflorum* is found only in Thailand, Malaysia and the Philippines [1].
Synonyms with the epithets *sessiliflorum* and *cuneadenum* belong to the var. *sessiliflorum* [1].

A. conspicuum (Craib)I.Nielsen [1]

Pithecellobium conspicuum Craib [1].
Not climbing[1]; Tree[1]; Perennial[1].
Indo-China: Thailand(N) [1]. Asia: Burma(N) [1].
Description[1]; Distribution Map[8].
This species is not yet known with fully ripe pods. The unripe pod suggests an affinity to the Indo-Chinese species with cylindrical straight pods [1].

A. contortum (Mart.)I.Nielsen [1]

Abarema contorta (Mart.)Kosterm. [1]; *Feuilleea contorta* (Mart.)Kuntze [1]; *Pithecellobium contortum* Mart. [1].
Not climbing[1]; Tree[1]; Perennial[1].
Indo-China: Thailand(N) [1]. Asia: Indonesia-ISO(N) [1]; Malaysia-ISO(N) [1]; Borneo[1].
Description[1]; Distribution Map[10]; Illustration[1].

A. cordifolium (T.L.Wu)I.Nielsen [10]
Archidendron sp. 20 I.Nielsen [2,10]; *Zygia cordifolia* T.L.Wu [10].
Indo-China: Vietnam(N) [10]. Asia: China(N) [10].
Description[2].

A. dalatense (Kosterm.)I.Nielsen [2]
Abarema dalatensis Kosterm. [2]; *Albizia dalatensis* (Kosterm.)Huang [10].
Not climbing[2]; Tree[2]; Perennial[2].
Indo-China: Vietnam(N) [2]; China(N) [147].
Description[2]; Distribution Map[8]; Illustration[2].

A. eberhardtii I.Nielsen [2]
Albizia eberhardtii (I.Nielsen)Huang [10]; *Cylindrokelupha macrophylla* T.L.Wu [10]; *Cylindrokelupha platyphylla* sensu Kosterm. [2,13].
Not climbing[2]; Tree[2]; Perennial[2].
Indo-China: Vietnam(N) [2]. Asia: China(N) [147].
Description[2]; Distribution Map[8]; Illustration[2].
Endemic to northern Vietnam [2].

A. ellipticum (Blanco)I.Nielsen [1]
Abarema elliptica (Blanco)Kosterm. [1]; *Albizzia fasciculata* (Benth.)Kurz,p.p. [1]; *Pithecellobium ellipticum* (Blanco)Hassk. [1]; *Pithecellobium fasciculatum* Benth. [1].
Not climbing[1]; Tree[1]; Perennial[1].
Indo-China: Thailand(N) [1]. Asia: Indonesia-ISO(N) [1]; Malaysia-ISO(N) [1]; Borneo[1]; Philippines(N) [1].
Description[1,10]; Distribution Map[10]; Illustration[1,10].
Only subsp. *ellipticum* occurs in Indo-China [10].

A. glomeriflorum (Kurz)I.Nielsen [1]
Abarema glomeriflora (Kurz)Kosterm. [1]; *Albizzia glomeriflora* Kurz [1]; *Pithecellobium glomeriflorum* (Kurz)Kurz [1].
Not climbing[1]; Shrub or tree[1]; Perennial[1].
Indo-China: Thailand(N) [1]. Asia: Burma(N) [1].
Description[1]; Distribution Map[8]; Illustration[1].

A. jiringa (Jack)I.Nielsen [1]
Albizia jiringa (Jack)Kurz [1]; *Pithecellobium jiringa* (Jack)Prain [1]; *Pithecellobium lobatum* Benth. [1]; *Zygia jiringa* (Jack)Kosterm. [1].
Not climbing[1]; Tree[1]; Perennial[1].
Indo-China: Thailand(N) [1]. Asia: Burma(N) [1]; Indonesia-ISO(N) [1]; Malaysia-ISO(N) [1]; Borneo[1].
Description[1]; Distribution Map[10]; Illustration[1].
Food or Drink[1].
Seeds are poisonous to the kidneys when raw. Often cultivated [1].

A. kerrii (Gagnep.)I.Nielsen [2]
Abarema kerrii (Gagnep.)Kosterm. [2]; *Abarema yunnanense* sensu auct. [2]; *Pithecellobium kerrii* Gagnep. [2].
Not climbing[2]; Shrub or tree[2]; Perennial[2].
Indo-China: Laos(N) [2]; Vietnam(N) [2]. Asia: China(N) [2].
Description[2]; Distribution Map[8]; Illustration[2].
A. yunnanense (Kosterm.)I.Nielsen may be conspecific [10].

A. laoticum (Gagnep.)I.Nielsen [2]
Abarema robinsonii sensu auctt. [2]; *Cylindrokelupha robinsonii* sensu Kosterm.,p.p. [2]; *Paralbizzia robinsonii* sensu auct. [[2];*Pithecellobium laoticum* Gagnep. [2].
Not climbing[2]; Tree[2]; Perennial[2].
Indo-China: Laos(N) [2]; Thailand(N) [2]; Vietnam(N) [2]. Asia: Burma(N) [2].
Description[2,8]; Distribution Map[8]; Illustration[2].

A. lucidum (Benth.)I.Nielsen [2]

Abarema lucida (Benth.)Kosterm. [2]; *Albizzia championii* Benth. [2]; *Pithecellobium lucidum* Benth. [2].
Not climbing[2]; Tree[2]; Perennial[2].
Indo-China: Cambodia(N) [2]; Laos(N) [2]; Thailand(N) [2]; Vietnam(N) [2]. Asia: China(N) [2]; Taiwan(N) [2].
Description[1,2]; Distribution Map[8]; Illustration[1,2].
Wood[2].

A. occultatum (Gagnep.)I.Nielsen [2]

Abarema occultata (Gagnep.)Kosterm. [2]; *Pithecellobium occultatum* Gagnep. [2].
Not climbing[2]; Tree[2]; Perennial[2].
Indo-China: Cambodia(N) [2]; Vietnam(N) [2].
Description[2]; Distribution Map[8]; Illustration[2].

A. pellitum (Gagnep.)I.Nielsen [2]

Abarema globosa sensu auct.,p.p. [2]; *Abarema pellita* (Gagnep.)Kosterm. [2]; *Pithecellobium pellitum* Gagnep. [2].
Not climbing[2]; Tree[2]; Perennial[2].
Indo-China: Laos(N) [2]; Vietnam(N) [2].
Description[2]; Distribution Map[8]; Illustration[2].

A. poilanei (Kosterm.)I.Nielsen [2]

Abarema poilanei Kosterm. [2]; *Cylindrokelupha annamense* Kosterm.,p.p. [2]; *Ortholobium annamense* Gagnep. [2]; *Pithecellobium corymbosum* sensu Gagnep, p.p. [2].
Not climbing[2]; Tree[2]; Perennial[2].
Indo-China: Vietnam(N) [2].
Description[2]; Distribution Map[8]; Illustration[2].
Species is known only in Vietnam [2].

A. quocense (Pierre)I.Nielsen [2]

Abarema quocense (Pierre)Kosterm.,p.p. [2]; *Pithecellobium jiringa* sensu Craib,p.p. [2]; *Pithecellobium quocense* Pierre [2].
Not climbing[2]; Tree[2]; Perennial[2].
Indo-China: Cambodia(N) [2]; Thailand(N) [2]; Vietnam(N) [2].
Description[1,2]; Distribution Map[8]; Illustration[1,2].
Food or Drink[2].
Very close to *A. jiringa* [1].

A. robinsonii (Gagnep.)I.Nielsen [2]

Abarema robinsonii (Gagnep.)Kosterm.,p.p. [2]; *Albizia robinsonii* (Gagnep.)Huang [10]; *Cylindrokelupha balansae* sensu Kosterm. [2]; *Cylindrokelupha platyphylla* Kosterm.,p.p. [2]; *Cylindrokelupha poilanei* Kosterm. [2]; *Cylindrokelupha robinsonii* (Gagnep.)Kosterm.,p.p. [2]; *Paralbizzia robinsonii* (Gagnep.)Kosterm.,p.p. [2]; *Pithecellobium corymbosum* sensu Gagnep. [2]; *Pithecellobium robinsonii* Gagnep. [2].
Not climbing[2]; Tree[2]; Perennial[2].
Indo-China: Vietnam(N) [2]. Asia: China(N) [147].
Description[2,8]; Distribution Map[8]; Illustration[2].

A. tetraphyllum (Gagnep.)I.Nielsen [2]

Abarema tetraphylla (Gagnep.)Kosterm. [2]; *Pithecellobium tetraphyllum* Gagnep. [2].
Not climbing[2]; Tree[2]; Perennial[2].
Indo-China: Vietnam(N) [2].
Description[2]; Distribution Map[8]; Illustration[2].
Known only from northern Vietnam [2].

A. tonkinense I.Nielsen [2]

Albizia tonkinensis (I.Nielsen)Huang [10].
Not climbing[2]; Tree[2]; Perennial[2].
Indo-China: Vietnam(N) [2]. Asia: China(N) [147].
Description[2,8]; Distribution Map[8]; Illustration[2,8].

A. turgidum (Merr.)I.Nielsen [2]

Albizzia croizatiana Metcalf [2]; *Albizzia turgida* (Merr.)Chun [2]; *Cylindrokelupha turgida* (Merr.)T.L.Wu [10]; *Paralbizzia turgida* (Merr.)Kosterm. [2]; *Pithecellobium turgidum* Merr. [2].
Not climbing[2]; Tree[2]; Perennial[2].
Asia: China(N) [2]. Indo-China: Vietnam(N) [2].
Description[2]; Illustration[2].

A. utile (Chun & How)I.Nielsen [2]

Abarema utile (Chun & How)Kosterm. [2]; *Pithecellobium angulatum* sensu Merr. [2,12]; *Pithecellobium utile* Chun & How [2].
Not climbing[2]; Shrub[2]; Perennial[2].
Asia: China(N) [2]. Indo-China: Vietnam(N) [2].
Description[2]; Distribution Map[8]; Illustration[2].
This species is very close to *A. glomeriflorum*. Further material may show a cline between the two species [8].

A. sp. I.Nielsen [10]

Archidendron sp.1 I.Nielsen [8]; *Archidendron* sp.19 I.Nielsen [2].
Indo-China: Vietnam(N) [2,8].
Description[2,8]; Illustration[2].

CALLIANDRA Benth.

A genus of about 200 species of trees and shrubs. Mainly in tropical America but with a few species in Africa, India and SE Asia.

C. sp. I.Nielsen [5] (Provisional)

Not climbing[5]; Tree[5]; Perennial[5] .
Indo-China: Vietnam(N) [5].
Description [5].
A new species, but insufficiently known for description.
Based on *Poilane* 9150.

CATHORMION (Benth.)Hassk.

Tropical forest trees; the genus is usually regarded as being synonymous with *Albizia.*

C. umbellatum (Vahl)Kosterm. [2]

Feuilleea umbellata (Vahl)Kuntze [2]; *Inga concordiana* DC. [2]; *Mimosa umbellata* Vahl [2];*Pithecolobium malayanum* Pierre [2]; *Pithecolobium umbellatum* (Vahl)Benth. [2].
Not climbing[1]; Shrub or tree[1]; Perennial[1].
Indo-China: Cambodia(N) [2]; Laos(N) [1]; Thailand(N) [1]; Vietnam(N) [1]. Asia: India(N) [1]; Indonesia-ISO(N) [1]; Sri Lanka(N) [1].
Description[1,2]; Illustration[1,2].
About twelve tropical species occur. Only one species in tropical Asia [1].

PITHECELLOBIUM C. Mart.

A small mainly neotropical genus. One species is very widely planted in the tropics as a thorny hedge.

P. dulce (Roxb.)Benth. [2]

Inga dulcis (Roxb.)Willd. [2]; *Mimosa dulcis* Roxb. [2].
Not climbing[1]; Shrub or tree[1]; Perennial[1].
Indo-China: Cambodia(N) [2]; Laos(I) [2]; Thailand(I) [1]; Vietnam(I) [2].
Description[1,2]; Illustration[1,2].
Environmental[2]; Food or Drink[1].
Madras Thorn[2]; Manila Tamarind[2].
Originally introduced from central America. Now cultivated and naturalised in old clearings [1].

P. tenue Craib [1]

Acacia tenue (Craib)Kosterm. [1]; *Thailentadopsis tenuis* (Craib)Kosterm. [1].
Not climbing[1]; Shrub or tree[1]; Perennial[1].
Indo-China: Thailand(N) [1].
Description[1]; Distribution Map[8]; Illustration[1].
Endemic to Thailand [1].

P. vietnamense I.Nielsen [2]

Not climbing[2]; Shrub[2]; Perennial[2].
Indo-China: Vietnam(N) [2].
Description[2,8]; Distribution Map[8]; Illustration[2,8].
Endemic to southern Vietnam.
Generic position is provisional [2].

SAMANEA (DC.)Merr.

A neotropical genus of forest or woodland trees, sometimes included within *Albizia*. One species is very widely planted in the tropics as a shade tree; its fruits provide a valuable dry-season fodder.

S. saman (Jacq.)Merr. [2].

Albizzia saman (Jacq.)F.Muell; *Enterolobium saman* (Jacq.)Prain [2]; *Mimosa saman* Jacq. [2]; *Pithecolobium saman* (Jacq.)Benth. [2].
Not climbing[1]; Tree[1]; Perennial[1].
Indo-China: Cambodia(I) [2]; Laos(I) [2]; Thailand(I) [1]; Vietnam(I) [2].
Description[1,2]; Illustration[1,2].
Environmental[1]; Forage[1]; Wood[1].
East Indian Walnut[1]; Rain Tree[1].
18 species occur in Northern South America. 1 sp. introduced into Thailand [1].
Cultivated and naturalized throughout Indo-China [1].

SERIANTHES Benth.

About 10 species, all from southeast Asia and Malesia. some yield timber.

Serianthes grandiflora Benth. [1]

S. dilmyi Fosb. [1].
Not climbing[1]; Tree[1]; Perennial[1].
Indo-China: Thailand(N) [1]. Asia: Indonesia-ISO(N) [1]; Malaysia-ISO(N) [1]; Philippines(N) [1].
Description[1]; Illustration[1].

MIMOSEAE

ADENANTHERA L.

Trees, from tropical Asia and the Pacific. Seeds red, or red and black. Some species yield useful timber, and others are planted widely in the Old World tropics as street trees and for shade. A recent revision (Nielsen & Guinet 1992; ref.93) takes a narrower view of the species than has been customary and recognizes 12 species.

A. microsperma Teijsm. & Binn. [93]

A. pavonina var. *microsperma* (Teijsm. & Binn.)Nielsen, p.p. [93]
Not climbing[93]; Tree[93]; Perennial[93].
Indo-China: Cambodia(U) [93]; Laos(U) [93]; Thailand(U) [93]; Vietnam(U) [93]. Asia: Burma(U) [93]; China(U) [93]; Indonesia-ISO(U) [93]; Malaysia-ISO(U) [93]
Description[93]; Distribution Map[93]; Illustration[93].
Environmental[93].

A. pavonina L. [93]

Not climbing[93]; Tree[93]; Perennial[93].
Indo-China: Cambodia(N) [2]; Laos(N) [2]; Thailand(N) [1] Vietnam(N) [2]. Asia: Burma(N) [2]; China(N) [2]; India(N) [2]; Malaysia-ISO(N) [2] Sri Lanka(N) [2]. Australasia: Australia(N) [2].
Description[1,2,93]; Illustration[1,2,93].
Chemical Products[93]; Domestic[93]; Environmental[93]; Food or Drink[93]; Wood[93]

A. tamarindifolia Pierre [93]

A. pavonina var. *microsperma* (Teijsm. & Binn.)Nielsen, p.p. [93]
Not climbing[93]; Shrub or Tree[93]; Perennial[93].
Indo-China: Vietnam(N) [93].
Description[93]; Distribution Map[93]; Illustration[93].
Known only from the type and from material cultivated in the Saigon Botanic Garden.

DICHROSTACHYS (DC.)Wight & Arn.

Shrubs, usually spiny. There are about 12 species, mostly in Madagascar but with one very wide-ranging and variable species which has also been introduced into parts of the New World, where it can become a pest.

D. cinerea Wight & Arn. [1,15]

Not climbing[1]; Tree[1]; Perennial[1].
Indo-China: Thailand(I) [1].
Description[1].
Environmental[1].
A native of tropical Africa. Has been introduced into Thailand as an ornamental tree; not yet naturalized [1].

var. **malesiana** Brenan & Brummitt [15]

D. cinerea var. *paucijuga* Miq. [15].
Not climbing[1]; Tree[1]; Perennial[1].
Indo-China: Thailand(I) [1]. Asia: Indonesia-ISO(I) [15]. Australasia: Australia(I) [15].
Description[1,15].
A subspecies, *burmana* Brenan & Brummitt, is confined to Burma.

ENTADA Adans.

A genus of about 30 species in the tropics of both hemispheres. Some are small trees of seasonally dry fire-swept woodlands; others are lianas, sometimes very large. The difference in size between the flower and the fruit must be greater in some of these lianes than in any other plant.

E. glandulosa Gagnep. [2]

E. tamarindifolia Gagnep. [2].
Climbing[2]; Herb or Shrub[2]; Perennial[2] .
Indo-China: Cambodia(N) [1]; Laos(N) [1]; Thailand(N) [1]; Vietnam(N) [1]. Asia: Burma(N) [1].
Description[1,2]; Illustration[1,2].
Medicine[2].
According to some labels this species is said to be a shrub, to others a perennial herb [1].

E. phaseoloides (L.)Merr. [2]

E. rumphii Scheff. [2]; *E. scandens* (L.)Benth.,p.p. [2]; *E. tonkinensis* Gagnep. [2].
Climbing[2]; Shrub[2]; Perennial[2].
Indo-China: Thailand(I) [4]; Vietnam(N) [2]. Asia: China(N) [2]; Indonesia(N) [2]; Malaysia(N) [2]; Philippines(N) [2]. Australasia: Australia(N) [2].
Description[2]; Illustration[2].

E. reticulata Gagnep. [2]

E. tamarindifolia Gagnep.,p.p.
Climbing or not[2]; Shrub[2]; Perennial[2].
Indo-China: Cambodia(N) [2]; Laos(N) [2].
Description[2]; Illustration[2].

E. rheedii Spreng. [1]

Climbing[2]; Shrub[2]; Perennial[2].
Indo-China: Cambodia(N) [2]; Laos(N) [2]; Thailand(N) [1]; Vietnam(N) [2]. Asia: Burma(N) [1]; China(N) [1]; India(N) [1].
Description[1,2]; Illustration[1,2].
Domestic[2]; Food or Drink[2]; Medicine[2].

subsp. **rheedii** [1,2]

E.pursaetha DC. [1]; *E. monostachya* DC. [1]; *E. scheffleri* Ridl. [1].
Climbing[2]; Shrub[2]; Perennial[2].
Indo-China: Cambodia(N) [2]; Laos(N) [2]; Thailand(N) [1]; Vietnam(N) [2].
Description[1,2]; Illustration[1,2].
Domestic[2]; Food or Drink[2]; Medicine[2].

subsp. **sinohimalensis** (Grierson & Long)ined. [2]

E. laotica Gagnep. [2]; *E. pursaetha* subsp. *sinohimalensis* Grierson & Long [2].
Climbing[2]; Shrub[2]; Perennial[2].
Indo-China: Laos(N) [2]. Asia: Bangladesh(N) [2]; Burma(N) [2]; China(N) [2]; India(N) [2]; Nepal(N) [2].
Description[2]; Illustration[2].

E. spiralis Ridl. [1]

E. schefferi Ridl.,p.p. [2].
Climbing[1]; Shrub[1]; Perennial[1].
Indo-China: Thailand(N) [1]. Asia: Indonesia-ISO(N) [1]; Malaysia-ISO(N) [1].
Description[1]; Illustration[1].

LEUCAENA Benth.

A genus of some 40 species, all trees or shrubs, mostly native to Central America. One species has been very widely introduced throughout the tropics as a multipurpose tree, yielding timber, fuel, and fodder, *inter alia*.

L. leucocephala (Lam.)De Wit [2]

L. glauca (Willd.)Benth. [2].
Not climbing[1]; Shrub or Tree[1]; Perennial[1].
Indo-China: Cambodia(I) [2]; Laos(I) [2]; Thailand(I) [1]; Vietnam(I) [2].
Description[1,2]; Illustration[1,2].
Environmental[1]; Food or Drink[1]; Forage[1]; Toxins[1]; Wood[1].
Cultivated in Thailand [1]; Capable of colonising bare soil [1].

MIMOSA L.

A very large genus of trees, shrubs amd herbs, some scrambling or climbing. Most of the 470 species are native to tropical America; a few are widely introduced and are often noxious weeds. See R.C.Barneby (1991): Ref. 172.

M. diplotricha Sauvalle [172]

M. invisa Colla [172].
Indo-China: Laos(I) [2]; Thailand(I) [1]; Vietnam(I) [2].
Climbing or not[1]; Shrub[1]; Perennial[1].
Description[1,2]; Illustration[1,2].
Environmental[1]; Medicine[2]; Wood[2].
A native of Brazil introduced into the Old World as a cover crop.

M. pigra L. [2]

Indo-China: Thailand(I) [1]; Vietnam(I) [2].
Climbing or not[1]; Shrub[1]; Perennial[1].
Description[1,2]; Illustration[1,2].
Weed[1]; Wood[1].
Pantropical weed probably from South America, introduced all over the tropics [1].

M. pudica L. [2]

Indo-China: Cambodia(N) [2]; Laos(N) [2]; Thailand(N) [1]; Vietnam(N) [2]. Asia: Indonesia-ISO(N) [1].
Not climbing[1]; Herb[1]; Annual or Perennial[1].
Description[1,2]; Illustration[1,2].
Environmental[1]; Medicine[4]; Weed[1].
Poilane 717: Pantropical. Common as a weed all over Thailand.

var. **hispida** Brenan [2]

Not climbing[1]; Herb[1]; Annual or Perennial[1].
Indo-China: Cambodia(N) [2]; Laos(N) [2]; Thailand(N) [1]; Vietnam(N) [2]. Asia: Indonesia-ISO(N) [1].
Description[1,2]; Illustration[1,2].
Weed[1].
Pantropical. Common as a weed throughout Thailand, but poorly collected [1].

var. **unijuga** (Duchass. & Walp.)Griseb. [2]

M. unijuga Duchass. & Walp. [1].
Not climbing[1]; Herb[1]; Annual or Perennial[1].
Indo-China: Cambodia(N) [2]; Thailand(N) [1]; Vietnam(N) [2].
Description[1,2].
Weed[1].
Pantropical. Common as a weed throughout Thailand, but is poorly collected [1,2].

NEPTUNIA Lour.

Short-lived herbs, with about 12 species throughout the tropics. Some species are aquatics with floating stems.

N. javanica Miq. [2]

N. triquetra sensu auctt. [2].
Not climbing[2]; Herb[2]; Perennial[2].
Indo-China: Cambodia(N) [2]; Thailand(N) [2]; Vietnam(N) [4]. Asia: Burma(N) [2]; Indonesia-ISO(N) [2].
Description[1,2]; Illustration[1,2].
Close to *N. triquetra* (Vahl)Benth. from India. The name *N. acinaciformis* has been used, (Windler, Aust.J.Bot.14:393(1966)) but Nielsen [1] (1981,1985) considers this name ambiguous until the type of *Desmanthus acinaciformis* Spanoghe, the basionym of *N. acinaciformis* (Spanoghe)Miq. is located. Until then the name *N.javanica* has to be maintained [1].

N. oleracea Lour. [1]

N. natans (L.f.)Druce [2].
Neptunia prostrata (Lam.)Baill. [2].
Not climbing[1]; Herb[1]; Perennial[1].
Indo-China: Cambodia(N) [2]; Laos(N) [2]; Thailand(N) [1]; Vietnam(N) [1].
Description[1,2]; Illustration[1,2].
Food or Drink[1].
Occurs in the tropics of both hemispheres. Often cultivated [1]; wild plants are also protected for use as food [2].

XYLIA Benth.

13 species in the tropical Old World, most diverse in Africa, mostly large or small trees. Some yield valuable timber.

X. xylocarpa (Roxb.)Taubert [2]

X. dolabriformis Benth. [2]; *Acacia xylocarpa* (Roxb.)Willd. [2]; *Inga xylocarpa* (Roxb.)DC. [2]; *Mimosa xylocarpa* Roxb. [2].
Not climbing[1]; Tree[1]; Perennial[1].
Indo-China: Cambodia(N) [1]; Laos(N) [1]; Thailand(N) [1]; Vietnam(N) [1]. Asa: Burma(N) [1].
Description[1,2]; Illustration[1,2].
Domestic[2]; Medicine[1]; Wood[2].
Two varieties occur, *xylocarpa* & *kerrii*; only the latter exists in Indo-China [1].

var. **xylocarpa** [1]

Not climbing[1]; Tree[1]; Perennial[1].
Asia: Burma(N) [1]; India(N) [1].
Description[1,2].

var. **kerrii** (Craib & Hutch.)I.Nielsen [1]

X. kerrii Craib & Hutch. [1].
Indo-China: Cambodia(N) [1]; Laos(N) [1]; Thailand(N) [1]; Vietnam(N) [1]. Asia: Burma(N) [1].
Not climbing[1]; Tree[1]; Perennial[1].
Description[1,2]; Illustration[1,2].
Domestic[2]; Medicine[1]; Wood[2].
Occurs in fire-influenced forests [1].

PARKIEAE

PARKIA R.Br.

About 40 species, all trees, throughout the tropical regions. Most diverse in South America. The southeast Asian and African taxa are relatively closely related.

P. leiophylla Kurz [1]

Indo-China: Thailand(N)[1]. Asia: Burma(N)[1]; China(N)[141].
Not climbing[1]; Tree[1]; Perennial[1].
Description[1]; Illustration[1].
Food or Drink[1].
Chinese material is not typical [141].

P. speciosa Hassk. [1]

Indo-China: Thailand(N)[1]. Asia: Indonesia-ISO(N)[1]; Malaysia-ISO(N)[1]; Philippines(N) [141].
Not climbing[1]; Tree[1]; Perennial[1].
Description[1]; Illustration[1].
Food or Drink[1].
Pete, Petai [141].
Also cultivated[1].

P. sumatrana Miq. [2]

P. insignis Kurz [2].
Not climbing [2]; Tree [2]; Perennial [2].
Indo-China: Cambodia(N) [142]; Laos(N) [142]; Thailand(N) [142]; Vietnam(N) [142]. Asia: Burma(N) [142]; Indonesia-ISO(N) [142]; Malaysia-ISO(N) [142].
Description[1,2]; Illustration[1,2].
Domestic[2]; Food or Drink[2]; Wood[2].

subsp. **sumatrana** [142]

P. macrocarpa Miq. [142].
Not climbing [142]; Tree [142]; Perennial [142].
Asia: Indonesia-ISO(N) [142]; Malaysia-ISO(N) 142].
Description [142].

subsp. **streptocarpa** (Hance) H.C.F.Hopkins [142]

P. dongnaiensis Pierre [142]; *P. streptocarpa* Hance [142].
Not climbing [142]; Tree [142]; Perennial [142].
Indo-China: Cambodia(N) [142]; Laos(N) [142]; Thailand(N) [142]; Vietnam(N) [142]. Asia: Burma(N) [142]; Malaysia-ISO(N) [142].
Description [142].

P. timoriana (DC.)Merr. [1]

P. javanica sensu auctt.,p.p. [1]; *P. roxburghii* G.Don [1]; *Inga timoriana* DC. [1].
Not climbing[1]; Tree[1]; Perennial[1].
Indo-China: Thailand(N)[1]. Asia: Bangladesh(N) [141]; Burma(N) [141]; India(N)[1]; Indonesia-ISO(N)[1]; Malaysia-ISO(N)[1]; Philippines(N) [141].
Description[1]; Illustration[1].
Food or Drink[1].
Also cultivated; Nielsen regards *P. javanica* as a nomen dubium[1].

PAPILIONOIDEAE

ABREAE

ABRUS Adans.

An Old World genus of small woody climbers or woody herbs, all perennial. There has been some controversy over species delimitation in the genus.

A. longibracteatus J.-N.Labat [96]
Climbing [96]; Herb [96]; Perennial [96].
Indo-China: Laos(N) [96]; Vietnam(N) [96].
Description [96]; Illustration [96]
Apparently endemic to Indo-China.

A. precatorius L. [29]
Climbing [29]; Herb or Shrub[138]; Perennial [29].
Indo-China: Cambodia(N) [29]; Laos(N) [29]; Thailand(N) [47]; Vietnam(N) [29]. Asia: Burma(N) [138]; China(N) [138]; India(N) [138]; Indonesia-ISO(N) [138]; Malaysia-ISO(N) [138]; Philippines(N) [138]; Sri Lanka(N) [138].
Description [29]; Illustration [29].
Domestic [29]; Food or Drink [29]; Medicine [47]
Widespread in the Old World. Typical subspecies in Asia; another in Africa [138].
Widely cultivated and naturalised in the New World; almost always the African subspecies [138].

subsp. **precatorius** [138]
Climbing [138]; Herb or Shrub [138]; Perennial [138].
Indo-China: Cambodia(N) [138]; Laos(N) [138]; Thailand(N) [138]; Vietnam(N) [138]. Asia: Burma(N) [138]; China(N) [138]; India(N) [138]; Indonesia-ISO(N) [138]; Malaysia-ISO(N) [138]; Philippines(N) [138].
Description [138].

A. pulchellus Thwaites [29]
Climbing [29]; Shrub [29]; Perennial [29].
Indo-China: Cambodia(N) [29]; Laos(N) [29]; Thailand(N) [29]; Vietnam(N) [29]. Asia: China(N) [29]; Indonesia-ISO(N) [29].
Description [29]; Illustration [29].
Pantropical [29]; five subspecies of which three in Indochina [138].

subsp. **pulchellus** [29]
A. melanospermus Hassk. [29]
Indo-China: Cambodia(N) [29]; Laos(N) [29]; Thailand(N) [138]; Vietnam(N) [29]. Asia: Burma(N) [47]; China(N) [138]; India(N) [47]; Malaysia-ISO(N) [138]; Philippines(N) [47]; Sri Lanka(N) [138].
Climbing [29]; Shrub [29]; Perennial [29].
Description [29]; Illustration [29].
Mainly Asian in distribution [138].

subsp. **cantoniensis** (Hance)Verdc. [29]
A. cantoniensis Hance [29]; *A. cantoniensis* var. *hossei* Craib [29].
Climbing [29]; Shrub [29]; Perennial [29].
Indo-China: Thailand(N) [29]; Vietnam(N) [29]. Asia: China(N) [29].
Description [29].

subsp. **mollis** (Hance)Verdc. [29]

A. mollis Hance [29].
Climbing [29]; Shrub [29]; Perennial [29].
Indo-China: Cambodia(N) [29]; Laos(N) [29]; Thailand(N) [29]; Vietnam(N) [29]. Asia: China(N) [29]; Indonesia-ISO(N) [29]; Malaysia-ISO(N) [138].
Description [29]; Illustration [29].

AESCHYNOMENEAE

AESCHYNOMENE L.

A pantropical genus of about 150 species of herbs or softly woody shrubs, many of them growing in wet places. Some of the swamp species bear stem nodules.

A. americana L. [29]

A. javanica Miq. [29].
Not climbing[29]; Herb or Shrub[50]; Annual or Perennial[50].
Indo-China: Thailand(I)[50]; Vietnam(I)[50]. Asia: Indonesia-ISO(I)[50]; Philippines(I) [50].
Description[29]; Illustration[29].
Native to tropical and subtropical America [50].
Ubolchalaket s.n. 'Thailand' [4].

var. **glandulosa** (Poir.)Rudd [4,50]

Indo-China: Thailand(I)[4]. Asia: Indonesia-ISO(I)[50].
Not climbing[4]; Herb[4], Shrub; Perennial[4].
Description[50].
T.Smitinand 7950 'Thailand; erect herb or undershrub; introduced' [4].
In later work Rudd [170] questions the usefulness of recognizing infraspecific taxa as the species is so variable.

A. aspera L. [29]

A. trachyloba Miq. [29].
Not climbing[29]; Herb[29]; Perennial[29].
Indo-China: Cambodia(N)[29]; Laos(N)[29]; Thailand(N)[29]; Vietnam(N)[29]. Asia: India(N) [29]; Sri Lanka(N)[29].
Description[29]; Illustration[29].
Domestic [170].

A. indica L. [29]

Not climbing[29]; Herb[29]; Annual[29].
Indo-China: Cambodia(N)[29]; Laos(N)[29]; Thailand(N)[29]; Vietnam(N)[29] Asia: Burma(N) [29]; China(N)[29]; India(N)[29]; Japan(N)[29]; Sri Lanka(N)[29]; Taiwan(N)[29].
Description[29]; Illustration[29].
Environmental [170]; Domestic [170].

ARACHIS L.

A genus of about 20 herbaceous species. South American, but one species is cultivated throughout the tropics for its edible seeds. Probably grown in most tropical countries but under-recorded like most cultivated plants.

A. hypogaea L. [29]

Not climbing[29]; Herb[29]; Annual[29].
Asia: Laos(I)[29]; Vietnam(I)[29].
Description[29]; Illustration[29].
Food or Drink[29].
Probably originating in Brazil; widely cultivated in tropical regions [29].

CYCLOCARPA Baker

A single annual herbaceous species, widespread in the Old World tropics.

C. stellaris Urban [29]

Not climbing[29]; Herb[29]; Perennial[29].
Indo-China: Cambodia(N)[29]; Laos(N)[29]; Thailand(N)[4]; Vietnam(N)[29]. Asia: Malaysia-ISO(N)[29], Borneo[29].
Description[29]; Illustration[29].
A. Marcan 2569: 'Thailand' (K) [4].

GEISSASPIS Wight & Arn.

A genus of about 3 species, herbaceous, confined to tropical Asia. African taxa formerly placed here have been transferred to other genera, mainly *Humularia*.

G. cristata Wight & Arn. [29]

Indo-China: Cambodia(N)[29]; Thailand(N)[29]; Vietnam(N)[29]. Asia: Burma(N)[29]; China(N)[29]; India(N)[29].
Not climbing[29]; Herb[29]; Perennial[29].
Description[29]; Illustration[29].

ORMOCARPUM P.Beauv.

A genus of about 20 species of shrubs and small trees of the Old World tropics, occurring in both wet and dry regions.

O. cochinchinense (Lour.)Merr. [29]

Diphaca cochinchinensis Lour. [50]; *O. glabrum* Teijsm. & Binn. [29]; *O. orientale* (Spreng.)Merrill [50].
Indo-China: Thailand(N)[29]; Vietnam(N)[29] Asia: China(N)[29]; India(N)[29]; Taiwan(N)[29].
Not climbing[29]; Shrub[29]; Perennial[29].
Description[29,30]; Illustration[29,30].
Medicine[29].

SMITHIA Aiton

About 30 species of herbs and shrubs; Old World tropics but mainly in south Asia and Madagascar.

S. blanda Wall. [29]

S. blanda var. *paniculata* (Arn.)Baker [50]; *S. blanda* var. *racemosa* (Wight & Arn.)Baker [50]; *S. paniculata* Arn. [5]; *S. racemosa* Wight & Arn. [50]; *S. yunnanensis* Franchet [29].
Indo-China: Laos(N)[29]; Thailand(N)[4]. Asia: China(N)[29]; India(N)[29].
Not climbing[29]; Herb[50]; Annual or Perennial[50].
Description[29]; Illustration[29].
C. Charoenphol et al. 4848 'Thailand' (K)[4].

S. ciliata Royle [29]

Indo-China: Thailand(N)[29]; Vietnam(N)[29] Asia: China(N)[29]; India(N)[29]; Philippines(N) [29].
Not climbing[29]; Herb[29]; Annual[50].
Description[29,30]; Illustration[29].

S. conferta Smith [29]

S. conferta Smith var. *geminiflora* (Roth)Cooke [50]; *S. geminiflora* Roth [50]; *S. geminiflora* var. *conferta* (Smith)Baker [29].
Asia: India(N)[29]; Indonesia-ISO(N)[29]. Indo-China: Vietnam(N)[29].
Not climbing[29]; Herb[29]; Annual[29].
Description[29,30]; Illustration[29].

S. finetii Gagnep. [29]

Asia: Vietnam(N)[29].
Not climbing[29]; Herb[29],Shrub; Perennial[29].
Description[29]; Illustration[30].

S. sensitiva Aiton [29]

S. javanica Benth. [29].
Indo-China: Laos(N)[29]; Thailand(N)[29]; Vietnam(N)[29]. Asia: China(N)[29]; India(N)[29]; Indonesia-ISO(N)[29]; Philippines(N)[29]; Taiwan(N)[29].
Not climbing[29]; Herb[50]; Annual or Perennial[50].
Description[29,30]; Illustration[29,30].

ZORNIA J.Gmelin

A pantropical genus of small annual or perennial herbs. Specific limits are controversial but there may be as many as 70 species.

Z. cantoniensis Mohl. [29]

Z. gibbosa var. *cantoniensis* (Mohl.)Ohashi [29].
Indo-China: Vietnam(N)[29]. Asia: China(N)[29]; Indonesia-ISO(N)[29]; Taiwan(N)[29].
Not climbing[29]; Herb[29]; Perennial[29].
Description[29]; Illustration[29].

Z. gibbosa Spanoghe [29]

Zornia diphylla sensu auctt. [29,30]; *Zornia graminea* Spanoghe [29].
Indo-China: Thailand(N)[29]; Vietnam(N)[29]. Asia: Burma(N)[29]; China(N)[29]; Taiwan(N)[29].
Not climbing[29]; Herb[29]; Annual[29].
Description[29]; Illustration[29].

CROTALARIEAE

CROTALARIA

A genus of about 600 species of herbs, sometimes woody, and shrubs. Most numerous and diverse in Africa (over 500 species: see Polhill, R.M. (1982) Crotalaria in Africa and Madagascar), but pantropical. Some are toxic.

C. acicularis Benth. [29]

Not climbing[29]; Herb[29]; Annual[29].
Indo-China: Cambodia(N) [29]; Laos(N) [29]; Thailand(N) [48]; Vietnam(N) [29]. Asia: Burma(N) [29]; India(N) [29]; Taiwan(N) [29].
Description[29,48]; Distribution Map[48]; Illustration[48].
Medicine[29].

C. alata D.Don [29]

C. alata Leveille [29].
Not climbing[29]; Herb or Shrub[29]; Perennial[29].
Indo-China: Cambodia(N) [[29]; Laos(N) [29]; Thailand(N) [48]; Vietnam(N) [29]. Asia: China(N) [29].
Description[29,48].

C. albida Roth [29]

C. deflexa Benth. [29]; *C. hossei* Craib [29].
Not climbing[29]; Herb[29]; Annual or Perennial[29].
Indo-China: Cambodia(N) [29]; Laos(N) [29]; Thailand(N) [48]; Vietnam(N) [29]. Asia: China(N) [29]; India(N) [29]; Malaysia-ISO(N) [29]; Philippines(N) [29]; Taiwan(N) [29].
Description[29,48]; Distribution Map[48]; Illustration[48].
Domestic[47].

C. angustifolia (Gagnep.)Niyomdham [29]

C. linifolia var. *angustifolia* Gagnep. [29].
Not climbing[29]; Herb[29]; Annual[29].
Indo-China: Cambodia(N) [29]; Laos(N) [29]; Thailand(N) [29]; Vietnam(N) [29]. Asia: Philippines(N) [29].
Description[29,48]; Illustration[48].

C. annamensis Dy Phon [29].

Not climbing[29]; Herb[29]; Perennial[29].
Indo-China: Vietnam(N) [29].
Description[29]; Illustration[29].
Endemic to Vietnam. Close to *C.montana* & *C.umbellata* [29].

C. assamica Benth. [29]

Not climbing[29]; Shrub[29]; Perennial[29].
Indo-China: Laos(N) [29]; Thailand(N) [29]; Vietnam(N) [29]. Asia: China(N) [29]; India(N) [29]; Philippines(N) [29].
Description[29,48]; Distribution Map[48].
Environmental[29]; Medicine[29].

C. bracteata DC. [29]

Not climbing[29]; Shrub[29]; Perennial[29].
Indo-China: Cambodia(N) [29]; Laos(N) [29]; Thailand(N) [29]; Vietnam(N) [29]. Asia: Burma(N) [29]; China(N) [29]; Philippines(N) [29].
Description[29,48]; Distribution Map[48]; Illustration[48].
Medicine[29].

C. calycina Schrank [29]

Not climbing[29]; Herb[29]; Annual[29].
Indo-China: Cambodia(N) [29]; Laos(N) [29]; Thailand(N) [48]; Vietnam(N) [29].
Description[29,48].

C. cambodiensis Dy Phon [29]]

Not climbing[29]; Herb[29]; Perennial[29].
Indo-China: Cambodia(N) [29]; Laos(N) [29].
Description[29]; Illustration[29].

C. chinensis L. [29]

Not climbing[29]; Herb or Shrub[29].
Indo-China: Cambodia(N) [29]; Laos(N) [29]; Thailand(N) [48]; Vietnam(N) [29]. Asia: China(N) [29]; India(N) [29]; Indonesia-ISO(N) [48]; Philippines(N) [29]; Taiwan(N) [29].
Perennial[29].
Description[29,48]; Distribution Map[48]; Illustration[48].
Niyomdham agrees with Craib(1928) that *C.chinensis* is variable. It is often confused with *C.hirta* Willd.[48].

C. cleomifolia Bak. [29]

Not climbing[29]; Shrub[29]; Perennial[29].
Indo-China: Vietnam(I) [29].
Description[29].
Species introduced from Africa [29].

C. cytisoides DC. [48]

C. zsemaensis Gagnep. [48].
Not climbing[48]; Shrub[48]; Perennial[48].
Indo-China: Thailand(N) [48]. Asia: Burma(N) [48]; China(N) [48]; India(N) [48]; Nepal(N) [48].
Description[48]; Distribution Map[48]; Illustration[48].

C. dubia Benth. [48]
Not climbing[48]; Herb[48]; Annual[48].
Indo-China: Thailand(N) [48]. Asia: Burma(N) [48]; India(N) [48].
Description[48]; Distribution Map[48]; Illustration[48].

C. evolvuloides Benth. [29]
Not climbing[29]; Shrub[29]; Perennial[29].
Indo-China: Vietnam(N) [29]. Asia: India(N) [29]; Sri Lanka(N) [29].
Description[29].

C. ferruginea Benth. [29]
C. bodinieri Leveille [29]; *C. rufescens* Franchet [29].
Not climbing[29]; Herb[29]; Perennial[29].
Indo-China: Laos(N) [29]; Thailand(N) [29]; Vietnam(N) [29]. Asia: China(N) [29]; India(N) [29]; Indonesia-ISO(N) [48]; Malaysia-ISO(N) [48]; Nepal(N) [29]; Philippines(N) [29]; Sri Lanka(N) [29].
Description[29,48]; Distribution Map[48]; Illustration[48].
Rarely a climber [29].

C. filiformis Benth. [48]
Not climbing[48]; Herb[48]; Perennial[48].
Indo-China: Thailand(N) [48]. Asia: Burma(N) [48].
Description[48]; Illustration[48].

var. **filiformis** [48]
Not climbing[48]; Herb[48]; Annual[48].
Indo-China: Thailand(N) [48]. Asia: Burma(N) [48].
Description[48]; Illustration[48].

var. **kerrii** (Craib)Niyomdham [48]
C. kerrii Craib [48].
Not climbing[48]; Herb[48]; Annual[48].
Indo-China: Thailand(N) [48].
Description[48]; Illustration[48].
Known only in Thailand [48].

C. hirta Willd. [47]
Not climbing[47]; Herb[47]; Annual[47].
Indo-China: Cambodia(N) [47]; Vietnam(N) [47].
Description[47].

C. humifusa Benth. [4]
Not climbing[4]; Herb[4].
Indo-China: Thailand(N) [4].
Geesink et al. 8039: 'Thailand; prostrate herb' [4].

C. incana L. [29]
Not climbing[29]; Herb[29]; Annual[29].
Indo-China: Thailand(N) [48]; Vietnam(N) [29].
Description[29,48]; Illustration[29,48].
Pantropical[29].

C. juncea L. [29]
C. benghalensis Lam. [29]; *C. tenuifolia* Roxb. [29].
Not climbing[29]; Herb[29]; Annual[29].
Indo-China: Cambodia(N) [29]; Laos(N) [29]; Thailand(N) [48]; Vietnam(N) [29].
Description[29,48].
Environmental[29]; Fibre[29]; Food or Drink[29].
Pantropical[29].

C. kostermansii Niyomdham [48]
Not climbing[48]; Shrub[48]; Perennial[48].
Indo-China: Thailand(N) [48].
Description[48]; Distribution Map[48]; Illustration[48].
Only known from Thailand [48].

C. kurzii Kurz [29]

C. peguana Bak. [29].
Not climbing[29]; Herb[29]; Perennial[29].
Indo-China: Laos(N) [29]; Thailand(N) [29]; Vietnam(N) [29]. Asia: Burma(N) [29]; China(N) [29].
Description[29,48]; Distribution Map[48].

C. laburnifolia L. [48]

Not climbing[48]; Herb[48]; Perennial[48].
Indo-China: Thailand(N) [48]. Asia: India(N) [48]; Sri Lanka(N) [48]. Australasia: Australia(N) [48].
Description[48]; Illustration[48].
Weed[48].

C. larsenii Niyomdham [48]

Not climbing[48]; Herb[48]; Annual[48].
Indo-China: Thailand(N) [48].
Description[48]; Illustration[48].
Known only from Thailand [48].

C. medicaginea Lam. [29]

Not climbing[29]; Herb[29]; Perennial[29].
Indo-China: Laos(N) [29]; Thailand(N) [29]; Vietnam(N) [29]. Asia: China(N) 29]; India(N) [29]; Taiwan(N) [29].
Description[29,48]; Illustration[48].

var. **medicaginea** [29]

Not climbing[29]; Herb[29]; Perennial[29].
Indo-China: Laos(N) [29]; Thailand(N) [29]; Vietnam(N) [29]. Asia: China(N) [29]; India(N) [29]; Taiwan(N) [29].
Description[29,48]; Illustration[48].
In Thailand, var. *medicaginea* is often found along the sea coast and on river banks [48].

var. **neglecta** (Wight & Arn.)Bak. [29]

C. medicaginea var. *luxurians* (Benth.)Bak. [29]; *C. neglecta* Wight & Arn. [29]; *C. luxurians* Benth. [29].
Not climbing[29]; Herb[29]; Perennial[29].
Indo-China: Thailand(N) [29]; Vietnam(N) [29]. Asia: India(N) [29].
Description[29,48]; Illustration[48].
In Thailand, var. *neglecta* occupies arid areas inland [48].

C. melanocarpa Benth. [29]

Not climbing[29]; Herb[29]; Perennial[29].
Indo-China: Cambodia(N) [29]; Thailand(N) [29]; Vietnam(N) [29]. Asia: Burma(N) [29].
Description[29,48]; Distribution Map[48]; Illustration[29,48].

C. micans Link [169]

C. anagyroides Kunth [169]
Not climbing[29]; Shrub[29]; Perennial[29].
Indo-China: Laos(I) [29]; Thailand(I) [48]; Vietnam(I) [29]. Asia: Indonesia-ISO(I) [29].
Description[29,48]; Illustration[29].
Environmental[48].
Originally from S.America; cultivated & naturalised in tropical Asia [29].

C. montana Roth [29]

Not climbing[29]; Herb[29]; Annual[29].
Indo-China: Cambodia(N) [29]; Laos(N) [29]; Thailand(N) [29]; Vietnam(N) [29]. Asia: China(N) [29]; India(N) [29]; Philippines(N) [29].
Description[29,48]; Illustration[48].

var. **montana** [29]

C. linifolia sensu auctt. [29].
Not climbing[29]; Herb[29]; Annual[29].
Indo-China: Cambodia(N) [29]; Laos(N) [29]; Thailand(N) [48]; Vietnam(N) [29]. Asia: China(N) [29]; India(N) [29]; Philippines(N) [29].
Description[29,48]; Illustration[48].

C. nana Burm.f. [29]

Not climbing[29]; Herb[29]; Annual[29].
Indo-China: Cambodia(N) [29]; Thailand(N) [48]; Vietnam(N) [29]. Asia: India(N) [29]; Indonesia-ISO(N) [29]; Sri Lanka(N) [29].
Description[29]; Distribution Map[48]; Illustration[48].

C. neriifolia Benth. [29]

Not climbing[29]; Herb[29]; Perennial[29].
Indo-China: Laos(N) [29]; Thailand(N) [29]. Asia: Burma(N) [29]; India(N) [29].
Description[29,48]; Distribution Map[48].

C. occulta Benth. [29]

Not climbing[29]; Herb[29]; Perennial[29].
Indo-China: Laos(I) [29].
Description[29]; Illustration[29].
Originally from the Himalayan region [29].

C. pallida Aiton [29]

C. mucronata Desv. [29]; *C. siamica* Williams [29]; *C. striata* DC. [29].
Not climbing[29]; Herb[29]; Annual[29].
Indo-China: Cambodia(N) [29]; Laos(N) [29]; Thailand(N) [48]; Vietnam(N) [29].
Description[29,48]; Illustration[48].
Food or Drink[29]; Medicine[29]; Weed[48].
Pantropical[29].

C. phyllostachya Gagnep. [29]

Not climbing[29]; Herb[29]; Annual or Perennial[29].
Indo-China: Cambodia(N) [29]; Laos(N) [29]; Vietnam(N) [29].
Description[29].

C. prostrata Willd. [29]

Not climbing[29]; Herb[29]; Perennial[29].
Indo-China: Cambodia(N) [29]; Thailand(N) [29]; Vietnam(N) [29]. Asia: India(N) [29]; Indonesia-ISO(N) [29]; Pakistan(N) [29]; Sri Lanka(N) [29].
Description[29]; Illustration[29].

C. quinquefolia L. [29]

Not climbing[29]; Herb[29]; Annual[29].
Asia: India(N) [29]; Philippines(N) [29]. Indo-China: Cambodia(N) [29]; Laos(N) [29]; Thailand(N) [29]; Vietnam(N) [29].
Description[29,48]; Distribution Map[48].
Environmental[29]; Medicine[29]; Weed[48].

C. ramosissima Roxb. [29]

C. tomentosa Rottl. [29].
Not climbing[29]; Shrub[29]; Perennial[29].
Indo-China: Vietnam(I) [29]. Asia: India(N) [29].
Description[29]; Illustration[29].
Originally from India. Cultivated and naturalised in Vietnam [29].

C. retusa L. [29]

Not climbing[29]; Herb[29]; Annual[29].
Indo-China: Cambodia(N) [29]; Laos(N) [29]; Thailand(N) [48]; Vietnam(N) [29].
Description[29,48].
Food or Drink[29]; Medicine[29].

C. sessiliflora L. [29]

Not climbing[29]; Herb[29]; Annual[48].
Indo-China: Cambodia(N) [29]; Laos(N) [29]; Thailand(N) [48]; Vietnam(N) [29]. Asia: China(N) [29]; India(N) [29]; Japan(N) [29]; Philippines(N) [29]; Taiwan(N) [29].
Description[29,48].

C. shanica Lace [48]

Not climbing[48]; Herb[48]; Annual[48].
Indo-China: Thailand(N) [48]. Asia: Burma(N) [48].
Description[48]; Distribution Map[48]; Illustration[48].

C. spectabilis Roth [48]

Not climbing[48]; Herb[48]; Perennial[48].
Indo-China: Thailand(N) [48]. Asia: Burma(N) [48]; India(N) [48]; Nepal(N) [48].
Description[48]; Distribution Map[48]; Illustration[48].

subsp. **spectabilis** [48]

C. sericea Retz. [48].
Not climbing[48]; Herb[48]; Annual[48].
Asia: India(N) [48].
Description[48].
Pantropical, but the typical form has not been recorded from Thailand [48].

subsp. **parvibracteata** Niyomdham [48]

Not climbing[48]; Herb[48]; Perennial[48].
Indo-China: Thailand(N) [48]. Asia: Burma(N) [48]; Nepal(N) [48].
Description[48]; Distribution Map[48]; Illustration[48].

C. tetragona Andrews [29]

C. esquirolii Leveille [29].
Not climbing[29]; Shrub[29]; Perennial[29].
Indo-China: Laos(N) [29]; Thailand(N) [29]; Vietnam(N) [29]. Asia: China(N) [29]; India(N) [29].
Description[29,48]; Distribution Map[48].
Medicine[29].

C. trichotoma Bojer [170]

C. usaramoensis Bak. [29]; *C. zanzibarica* Benth. [170].
Not climbing[29]; Herb[29]; Annual or Perennial[29].
Indo-China: Vietnam(I) [29]. Asia: Taiwan(I) [29].
Description[29]; Illustration[29].
Originally from East Africa; introduced & naturalised in Vietnam [29].

C. ubonensis Dy Phon [49]

Not climbing[49]; Herb[49].
Indo-China: Thailand(N) [49].
Description[49]; Illustration[49].

C. umbellata Wight & Arn. [29]

Not climbing[29]; Herb[29]; Perennial[29].
Indo-China: Thailand(N) [29]; Vietnam(N) [29]. Asia: India(N) [29]; Sri Lanka(N) [29].
Description[29,48]; Distribution Map[48]; Illustration[29,48].

C. uncinella Lam. [29,48]

Not climbing[29,48]; Herb or Shrub[29,48]; Annual or Perennial[29,48].
Indo-China: Thailand(N) [29]; Vietnam(N) [29]. Asia: China(N) [29]; India(N) [29]; Taiwan(N) [29].
Description[29,48]; Distribution Map[48]; Illustration[48].
Subsp. *uncinella* is found in tropical Africa, Madagascar and the Mascarenes [29].

subsp. **elliptica** (Roxb.)Polhill [29]

C. elliptica Roxb. [29].
Not climbing[29]; Herb[29]; Annual or Perennial[29].
Indo-China: Thailand(N) [29]; Vietnam(N) [29]. Asia: China(N) [29]; India(N) [29]; Taiwan(N) [29].
Description[29,48]; Distribution Map[48]; Illustration[48].

C. valetonii Backer [29]

Not climbing[29]; Shrub[29]; Perennial[29].
Indo-China: Vietnam(I) [29]. Asia: Indonesia-ISO(N) [29].
Description[29]; Illustration[29].
Originally from Indonesia (Madura), this species is cultivated in the Saigon Botanic Gardens [29].

C. verrucosa L. [29]

Not climbing[29]; Herb[29]; Annual[29].
Indo-China: Cambodia(N) [29]; Laos(N) [29]; Thailand(N) [48]; Vietnam(N) [29].
Description[29,48].
Medicine[29].

ROTHIA Pers.

Two species, one in Africa and one in southern Asia.

R. indica (L.)Thuan [29]

R. trifoliata (Roth)Pers. [29,47]; *Dillwynia trifoliata* Roth [29]; *Trigonella indica* L. [29].
Not climbing[29]; Herb[29]; Annual[47].
Indo-China: Laos(N) [29]; Vietnam(N) [29]. Asia: India(N) [29]; Indonesia-ISO(N) [29]; Sri Lanka(N) [29].
Description[29]; Illustration[29].

DALBERGIEAE

DALBERGIA L.f.

A tropical genus of about 100 species of trees, shrubs and lianas. As in many other parts of the world, the Indo-Chinese taxa are much in need of revision; good material of the woody lianas is often scarce. Some of the tree species produce valuable and decorative timbers.

D. abbreviata Craib [90]

Climbing or not[90]; Shrub[90]; Perennial[90].
Indo-China: Thailand(N)[90].
Description[90].

D. balansae Prain [47].

D. lanceolaria Hemsley [47].
Not climbing[47]; Tree[47]; Perennial[47].
Indo-China: Vietnam(N)[47]. Asia: China(N)[47].
Description[47]; Illustration[80].

D. bariensis Pierre [47]

Not climbing[47]; Tree[47]; Perennial[47].
Indo-China: Thailand(N)[32]; Vietnam(N)[47].
Description[47]; Illustration[80].

D. boniana Gagnep. [47]

Not climbing[47]; Tree[47]; Perennial[47].
Indo-China: Vietnam(N)[47].
Description[47]; Illustration[47].

D. cambodiana Pierre [47]

Not climbing[47]; Tree[47]; Perennial[47].
Indo-China: Cambodia(N)[47]; Vietnam(N)[47].
Description[47]; Illustration[80].

D. cana Kurz [82]

Not climbing[82]; Tree[82]; Perennial[82].
Indo-China: Thailand(N)[32]. Asia: Burma(N)[82].
Description[47,82].

D. candenatensis (Dennst.)Prain [82]

D. monosperma Dalz. [82]; *D. torta* A.Gray [82]; *Drepanocarpus monospermus* Kurz [82].
Climbing[82]; Shrub[82]; Perennial[82].
Indo-China: Thailand(N)[32]; Vietnam(N)[47]. Asia: Bangladesh(N)[82]; Burma(N)[82]; China(N)[82]; India(N)[82]; Malaysia-ISO(N)[82]; Sri Lanka(N)[82]. Australasia: Australia(N) [82].
Description[47,82]; Illustration[80].
Chemical Products[82].

D. cochinchinensis Pierre [47]

Not climbing[47]; Tree[47]; Perennial[47].
Indo-China: Cambodia(N)[47]; Thailand(N)[47]; Vietnam(N)[47].
Description[47].

D. cultrata Benth. [82]

Not climbing[82]; Tree[82]; Perennial[82].
Indo-China: Laos(N)[47]; Thailand(N)[47]; Vietnam(N)[82]. Asia: Burma(N)[82].
Description[47,82]; Illustration[80].

var. **cultrata** [82]

D. fusca Prain [82].
Not climbing[82]; Tree[82]; Perennial[82].
Indo-China: Laos(N)[47]; Thailand(N)[47]; Vietnam(N)[82]. Asia: Burma(N)[82].
Description[47,82]; Illustration[80].
Two forms occur. Forma *fusca* occurs in Vietnam & Thailand. The typical form occurs in Laos & Thailand. Synonyms with epithet *fusca* refer to forma *fusca* [82].

var. **pallida** Craib [32]

Not climbing[82]; Tree[82]; Perennial[82].
Indo-China: Thailand(N)[32].
Description[32].

D. curtisii Prain [47]

D. discolor sensu Miq. [47]; *D. junghuhnii* Bak., p.p. [47].
Climbing or not[47]; Shrub[47]; Perennial[47].
Indo-China: Laos(N)[47]; Thailand(N)[32]; Vietnam(N)[47]. Asia: Indonesia-ISO(N)[47]; Malaysia-ISO(N)[47].
Description[47]; Illustration[80].

D. discolor Blume [47]

Climbing or not[4]; Shrub or Tree[4]; Perennial[47].
Indo-China: Thailand(N)[47]. Asia: Indonesia-ISO(N)[47]; Malaysia-ISO(N)[47].
Description[47]; Illustration[80].
Niyomdham et al. 1326: 'scandent shrub' [4].

D. dongnaiensis Pierre [47]

Not climbing[47]; Tree[47]; Perennial[47].
Indo-China: Cambodia(N)[47]; Thailand(N)[47]; Vietnam(N)[47].
Description[47]; Illustration[80].

D. duperreana Pierre [47]

Not climbing[47]; Tree[47]; Perennial[47].
Indo-China: Cambodia(N)[47]; Thailand(N)[32].
Description[47]; Illustration[80].

D. entadoides Prain [47]

D. foliacea Prain, p.p. [32].
Climbing[47]; Shrub or Tree[47]; Perennial[47].
Indo-China: Cambodia(N)[47]; Laos(N)[47]; Thailand(N)[32]; Vietnam(N)[47].
Description[47].

D. errans Craib [60]

Not climbing[60]; Tree[60]; Perennial[60].
Indo-China: Thailand(N)[60].
Description[60].

D. floribunda Craib [60]

Not climbing[60]; Tree[60]; Perennial[60].
Indo-China: Thailand(N)[60].
Description[60].

D. forbesii Prain [47]

D. parviflora sensu Prain, p.p. [47].
Not climbing[47]; Shrub[47]; Perennial[47].
Indo-China: Laos(N)[47]; Thailand(N)[32]. Asia: Indonesia-ISO(N)[47].
Description[47]; Illustration[80].

D. godefroyi Prain [47]

Climbing[47]; Shrub[47]; Perennial[47].
Indo-China: Cambodia(N)[47]; Laos(N)[47]; Thailand(N)[32].
Description[47]; Illustration[80].

D. hancei Benth. [4,88]

Climbing[88]; Shrub[88]; Perennial[88].
Indo-China: Thailand(N)[4]; Vietnam(N)[4]. Asia: Hong Kong(N)[88].
Description[88]; Illustration[80].
Kerr 20130: 'Thailand'; *W.T. Tsang* 29114: 'Vietnam' [4].

D. horrida (Dennst.)Mabb. [47,82]

Climbing[82]; Shrub[82]; Perennial[82].
Indo-China: Laos(N)[47]; Thailand(N)[32]; Vietnam(N)[47]. Asia: India(N)[82].
Description[47,82]; Illustration[80].

var. **horrida** [82]

D. multiflora Heyne [47,82]; *D. sympathetica* Nimmo [82].
Climbing[82]; Shrub[82]; Perennial[82].
Asia: India(N)[82].
Description[47,82]; Illustration[80].

var. **glabrescens** (Prain)Thoth. & Nair [47,82]

D. multiflora var. *glabrescens* Prain [82].
Climbing[82]; Shrub[82]; Perennial[82].
Indo-China: Laos(N)[47]; Thailand(N)[32]; Vietnam(N)[47]. Asia: India(N)[82].
Description[47,82].

D. hupeana Hance [81]

Not climbing[81]; Tree[81]; Perennial[81].
Indo-China: Laos(N)[47]; Vietnam(N)[47]. Asia: China(N)[81].
Description[47,81]; Illustration[80].

var. **hupeana** [81]

Not climbing[47]; Tree[47]; Perennial[47].
Indo-China: Laos(N)[47]; Vietnam(N)[47]. Asia: China(N)[81].
Description[81]; Illustration[80].

var. **laccifera** Eberh. & Dubard [47]

Not climbing[47]; Tree[47]; Perennial[47].
Indo-China: Laos(N)[47]; Vietnam(N)[47].
Description[47].
Chemical Products[47].

D. kerrii Craib [47]

Not climbing[47]; Tree[47]; Perennial[47].
Indo-China: Laos(N)[47]; Thailand(N)[47].
Description[47].

D. kurzii Prain [47,82]

Not climbing[82]; Tree[82]; Perennial[82].
Indo-China: Laos(N)[47]; Thailand(N)[32]. Asia: Burma(N)[82].
Description[82]; Illustration[80].

var. **kurzii** [47,82]

D. purpurea Kurz [47].
Not climbing[82]; Tree[82]; Perennial[82].
Indo-China: Laos(N)[47]. Asia: Burma(N)[82].
Description[82]; Illustration[80].

var. **truncata** Craib [32]

Not climbing[82]; Tree[82]; Perennial[82].
Indo-China: Thailand(N)[32].
Description[32].

D. lacei Thoth. [82]

Not climbing[82]; Tree[82]; Perennial[82].
Indo-China: Thailand(N)[32]. Asia: Burma(N)[82].
Description[82].

D. lakhonensis Gagnep. [47]

Not climbing[47]; Tree[47]; Perennial[47].
Indo-China: Laos(N)[47]; Thailand(N)[32].
Description[32,47]; Illustration[47].

var. **lakhonensis** [47]

Not climbing[47]; Tree[47]; Perennial[47].
Indo-China: Laos(N)[47].
Description[47]; Illustration[47].

var. **appendiculata** Craib [32,47]

Not climbing[47]; Tree[47]; Perennial[47].
Indo-China: Thailand(N)[32].
Description[32,47]; Illustration[47].

D. lanceolaria L.f. [82]

Not climbing[82]; Tree[82]; Perennial[82].
Indo-China: Cambodia(N)[47]; Laos(N)[47]; Thailand(N)[32]; Vietnam(N)[47]. Asia: Burma(N)[82]; India(N)[82]; Sri Lanka(N)[82].
Description[47,82]; Illustration[80].
Medicine[82]; Wood[82].

subsp. **lanceolaria** [82]

D. frondosa DC. [82].
Not climbing[82]; Tree[82]; Perennial[82].
Indo-China: Cambodia(N)[47]; Thailand(N)[4]; Vietnam(N)[4]. Asia: Burma(N)[82]; India(N)[82]; Sri Lanka(N)[82].
Description[47,82]; Illustration[80].
Medicine[82]; Wood[82].
Put 2810: 'Thailand'; *W.T.Tsang* 26995: 'Vietnam' [4].

subsp. **paniculata** (Roxb.)Thoth. [82]

D. hemsleyi Prain [82]; *D. maymensis* Craib [82]; *D. nigrescens* Kurz [82]; *D. paniculata* Roxb. [82].
Not climbing[82]; Tree[82]; Perennial[82].
Indo-China: Cambodia(N)[47]; Laos(N)[47]; Thailand(N)[32]; Vietnam(N)[47]. Asia: Burma(N)[82]; India(N)[82].
Description[47,82]; Illustration[80].
Wood[82].
The typical variety occurs in Thailand, Cambodia, Laos & Vietnam. Var. *hemsleyi* occurs in Cambodia, var. *saigonensis* occurs in Cambodia, Vietnam & Thailand [47] and var. *maymyensis* & var. *siamensis* occur in Thailand.[32].
Bunchuai 1592: '*D. nigrescens* var. *saigonensis* - Thailand' [4].

D. mammosa Pierre [47]

Not climbing[47]; Tree[47]; Perennial[47].
Indo-China: Vietnam(N)[47].
Description[47].

D. marcaniana Craib [89]

Not climbing[89]; Shrub[89]; Perennial[89].
Indo-China: Thailand(N)[89].
Description[89].

D. obtusifolia (Bak.)Prain [82]
D. glauca Kurz [82]; *D. ovata* var. *obtusifolia* Bak. [82].
Not climbing[82]; Tree[82]; Perennial[82].
Indo-China: Thailand(N)[4]. Asia: Burma(N) [82].
Description[82]; Illustration[80].
Winit 1864: 'Thailand' [4].

D. oliveri Prain [47,82]
D. laccifera Lanessan [47]; *D. prazeri* Prain [82].
Not climbing[82]; Tree[82]; Perennial[82].
Indo-China: Thailand(N)[47]; Vietnam(N)[47]. Asia: Burma(N)[82].
Description[47,82]; Illustration[80].
Wood[82].

D. ovata Benth. [47,82]
Not climbing[82]; Tree[82]; Perennial[82].
Indo-China: Thailand(N)[82]; Vietnam(N)[47]. Asia: Burma(N)[82].
Description[47,82]; Illustration[80].
Only the typical variety occurs in Indo-China[82].

var. **ovata** [82]
D. glauca Wall. [82].
Not climbing[82]; Tree[82]; Perennial[82].
Indo-China: Thailand(N)[82]; Vietnam(N)[47]. Asia: Burma(N)[82].
Description[47,82]; Illustration[80].

var. **glomeriflora** (Kurz)Thoth. [82]
D. glomeriflora Kurz [82].
Not climbing[82]; Tree[82]; Perennial[82].
Indo-China: Thailand(N)[4]. Asia: Burma(N)[82].
Description[82]; Illustration[80].
Kerr 10574: 'Thailand' [4].

D. parviflora Roxb. [82]
D. cumingiana Benth. [88]; *D. cumingii* Benth. [82]; *Drepanocarpus cumingii* Kurz [82].
Climbing[82]; Shrub[82]; Perennial[82].
Indo-China: Laos(N)[4]; Thailand(N)[32]. Asia: Burma(N)[82]; Indonesia-ISO(N)[82]; Malaysia-ISO(N)[82].
Description[82]; Illustration[80].
Kerr 21130: 'Laos' [4].

D. peguensis Thoth. [82]
D. oliveri sensu Prain, p.p. [80,82]; *D. paniculata* sensu Kurz, p.p. [82]; *D. purpurea* sensu Bak., p.p. [82].
Not climbing[82]; Tree[82]; Perennial[82].
Asia: Burma(N)[82].
Description[82].

D. pierreana Prain [47]
Climbing[47]; Shrub[47]; Perennial[47].
Indo-China: Cambodia(N)[47].
Description[47]; Illustration[80].

D. pinnata (Lour.)Prain [82]
D. tamarindifolia Roxb. [82].
Climbing or not[82]; Shrub or Tree[82]; Perennial[82].
Indo-China: Laos(N)[47]; Thailand(N)[47]; Vietnam(N)[82]. Asia: Burma(N)[47]; Indonesia-ISO(N)[47]; Malaysia-ISO(N)[47]; Philippines(N)[47].
Description[47,82].

D. pseudo-sissoo Miq. [82]
D. championii Thwaites [82]; *D. rostrata* Prain [82]; *D. sissoo* sensu Miq. [82].
Climbing or not[82]; Shrub[82]; Perennial[82].
Indo-China: Thailand(N)[4]. Asia: India(N)[82]; Malaysia-ISO(N)[82]; Sri Lanka(N)[82].
Description[82]; Illustration[82].
Haniff 360: 'Thailand' [4].

D. rimosa Roxb. [47,82]
Climbing or not[82]; Shrub[82]; Perennial[82].
Indo-China: Laos(N)[47]; Thailand(N)[32]; Vietnam(N)[47]. Asia: Burma(N)[82]; China(N)[82]; India(N)[82].
Description[47,82]; Illustration[80].

var. **rimosa** [82]
Climbing or not[82]; Shrub[82]; Perennial[82].
Indo-China: Laos(N)[47]; Thailand(N)[32]; Vietnam(N)[47]. Asia: Burma(N)[82]; China(N)[82]; India(N)[82].
Description[47,82]; Illustration[80].

var. **foliacea** (Benth.)Thoth. [82]
D. foliacea Benth. [82].
Climbing or not[82]; Shrub[82]; Perennial[82].
Indo-China: Thailand(N)[32]. Asia: Burma(N)[82].
Description[82]; Illustration[80].

D. stipulacea Roxb. [47]
Climbing or not[82]; Shrub[82]; Perennial[82].
Indo-China: Laos(N)[47]; Thailand(N)[32]; Vietnam(N)[82]. Asia: Burma(N)[82]; China(N)[82]; India(N)[82]; Nepal(N)[82].
Description[47,82]; Illustration[80].

D. succirubra Gagnep. & Craib [47]
Climbing[47]; Shrub[47]; Perennial[47].
Indo-China: Thailand(N)[47].
Description[47].

D. thorelii Gagnep. [47]
Not climbing[47]; Shrub[47]; Perennial[47].
Indo-China: Cambodia(N)[47]; Laos(N)[47]; Thailand(N)[4].
Description[47]; Illustration[47].
Smitinand 1163: 'Thailand' [4].

D. tonkinensis Prain [47]
Not climbing[47]; Tree[47]; Perennial[47].
Indo-China: Vietnam(N)[47].
Description[47]; Illustration[80].

D. velutina Benth. [82]
Climbing[82]; Shrub[82]; Perennial[82].
Indo-China: Laos(N)[47]; Thailand(N)[32]. Asia: Bangladesh(N)[82]; Burma(N)[82]; Malaysia-ISO(N)[82]; Borneo[82].
Description[47,82]; Illustration[80].

D. verrucosa Craib [60]
Climbing[60]; Shrub[60]; Perennial[60].
Indo-China: Thailand(N)[60].
Description[60].

D. volubilis Roxb. [47]
Climbing[47]; Shrub[47]; Perennial[47].
Indo-China: Laos(N)[47]; Thailand(N)[32]; Vietnam(N)[47]. Asia: Burma(N)[47]; India(N)[47].
Description[47,82]; Illustration[80].

var. **volubilis** [47]
D. confertiflora Benth., p.p. [47]; *D. purpurea* Benth., p.p. [47]; *D. stipulacea* sensu Gamble [47].
Climbing[47]; Shrub[47]; Perennial[47].
Asia: Burma(N)[47]; India(N)[47]. Indo-China: Laos(N)[47]; Thailand(N)[32].
Description[47]; Illustration[80].

var. **latifolia** Gagnep. [47]
Climbing[47]; Shrub[47]; Perennial[47].
Indo-China: Vietnam(N)[47].
Description[47].

PTEROCARPUS Jacq.

A genus of about 20 species, pantropical, all trees. Some produce valuable and decorative timbers, and red dyes are obtained from the wood of some species.

P. indicus Willd. [47]

P. wallichii Wight & Arn. [47]; *P.zollingeri* Miq. [47].
Not climbing[47]; Tree[47]; Perennial[47].
Indo-China: Cambodia(N)[47]; Thailand(N)[47]; Vietnam(N)[47]. Asia: Burma(U)[84]; India(N) [84]; Indonesia-ISO(N)[47]; Malaysia-ISO(N)[47]; Sri Lanka(N)[84]; Taiwan(I)[84].
Description[47,84]; Illustration[84].
Wood[47].

P. macrocarpus Kurz [84]

P. cambodianus (Pierre)Gagnep. [84]; *P. cambodianus* var. *calcicolus* Craib [84]; *P. glaucinus* Pierre [84]; *P. gracilis* Pierre [84]; *P. gracilis* var. *brevipes* Craib [84]; *P. gracilis* var. *nitidus* Craib [84]; *P. macrocarpus* var. *oblongus* (Pierre)Gagnep. [84]; *P. parvifolius* (Pierre)Craib [84]; *P. pedatus* (Pierre)Gagnep. [84].
Not climbing[84]; Tree[84]; Perennial[84].
Indo-China: Cambodia(N)[84]; Laos(N)[84]; Thailand(N)[84]; Vietnam(N)[84]. Asia: Burma(N) [84]; India(N)[84].
Description[84]; Illustration[84].

DESMODIEAE

The volume of 'Flore du Cambodge, du Laos et du Viet-nam' dealing with the Desmodieae is currently in preparation. Although Dr Vidal has annotated a draft of this scheme with various alterations, we have decided not to incorporate these in advance of publication of the Flore.

ALYSICARPUS Desv.

A genus of small herbs, annual or more usually perennial, occurring in the Old World tropics. Perhaps 25–30 species.

A. bupleurifolius (L.)DC. [29]

Hedysarum bupleurifolium L. [29]; *Hedysarum gramineum* Retz. [29].
Not climbing[29]; Herb[29]; Perennial[29].
Indo-China: Thailand(N) [29]; Vietnam(N) [29]. Asia: China(N) [29]; India(N) [[29]; Indonesia-ISO(N) [29]; Malaysia-ISO(N) [29]; Taiwan(N) [29].
Description[29]; Illustration[29].

A. rugosus (Willd.)DC. [29]

A. violaceus Schindler [29]; *A. wallichii* Wight & Arn. [29]; *Hedysarum rugosum* Willd. [29].
Indo-China: Laos(N) [29]; Thailand(N) [32]; Vietnam(N) [29].
Not climbing[29]; Herb[29]; Perennial[29].
Description[29]; Illustration[29].
Throughout tropical regions of the Old World [29].

A. vaginalis (L.)DC. [29]

A. nummularifolius (Willd.)DC. [29]; *A. vaginalis* var. *nummularifolius Miq.* [29]; *Hedysarum vaginale* L. [29].
Not climbing[29]; Herb[29]; Perennial[29].
Indo-China: Cambodia(N) [29]; Laos(N) [29]; Thailand(N) [32]; Vietnam(N) [29]. Asia: China(N) [30]; India(N) [30]; Indonesia-ISO(N) [30]; Philippines(N) [30].
Description[29]; Illustration[29].
Medicine[29].
Native of all tropical regions of the Old World. Introduced to tropical America [29].

CAMPYLOTROPIS Bunge

Herbs or shrubs; about 65 species, all confined to Asia.

C. bonii Schindler [29]

Lespedeza bonii (Schindler)Gagnep. [29].
Not climbing[29]; Shrub[29]; Perennial[29].
Indo-China: Vietnam [29].
Description[29]; Illustration[29].
Endemic to North Vietnam [29].

var. **angusticarpa** Schindler [32]

Lespedeza bonii var. *angusticarpa* (Schindler)Craib [32].
Not climbing[29]; Shrub[29]; Perennial[29].
Indo-China: Thailand(N) [32].
Description[42].

C. decora (Kurz)Schindler [32]

Lespedeza decora (Kurz)Craib [32].
Not climbing[29]; Shrub[29]; Perennial[29].
Asia: Burma(N) [32]. Indo-China: Thailand(N) [32].
Description[41].

C. harmsii Schindler [32]

Lespedeza harmsii Craib [32].
Not climbing[29]; Shrub[29]; Perennial[29].
Asia: China(N) [32]. Indo-China: Thailand(N) [32].
Description[40].

C. henryi Schindler [29]

Lespedeza henryi Schindler [29].
Not climbing[29]; Shrub[29]; Perennial[29].
Asia: China(N) [29]. Indo-China: Laos(N) [29]; Thailand(N) [29,32].
Description[29].
Found particularly on the banks of the Mekong River [29].

C. parviflora (Kurz)Schindler [29]

Lespedeza parviflora Kurz [29].
Not climbing[29]; Shrub[29]; Perennial[29].
Asia: Burma(N) [29]. Indo-China: Laos(N) [29]; Thailand(N) [29]; Vietnam(N) [29].
Description[29].

C. pinetorum (Kurz)Schindler [29]

Lespedeza pinetorum Kurz [29]; *Lespedeza tomentosa* Maxim. [29]; *Lespedeza velutina* Dunn [29].
Not climbing[29]; Shrub[29]; Perennial[29].
Indo-China: Laos(N) [29]; Thailand(N) [29]; Vietnam(N) [29]. Asia: Burma(N) [29]; China(N) [29].
Description[29].

C. splendens Schindler [29]

Lespedeza splendens (Schindler)Gagnep. [29].
Not climbing[29]; Shrub[29]; Perennial[29].
Indo-China: Vietnam(N) [29].
Description[29].
Endemic to southern Vietnam [29].

C. sulcata Schindler [32]

Lespedeza sulcata Craib [32].
Not climbing[29]; Shrub[29]; Perennial[29].
Asia: China(N) [32]. Indo-China: Thailand(N) [32].
Description[39].

CHRISTIA Moench.

Herbs or shrubs; about 10 species in tropical and subtropical Asia and Australasia.

C. constricta (Schindler)T.C.Chen [29]

Lourea constricta Schindler [29].
Not climbing[29]; Shrub[29]; Perennial[29].
Indo-China: Vietnam(N) [29].
Description[29]; Illustration[29].
Medicine[29].

C. convallaria (Schindler)Ohashi [29]

Lourea convallaria Schindler [29].
Not climbing[29]; Herb or Shrub[29]; Perennial[29].
Indo-China: Vietnam(N) [29].
Description[29]; Illustration[29].

C. lychnucha (Schindler)Ohashi [29]

Lourea lychnucha Schindler [29].
Not climbing[29]; Herb[29]; Perennial[29].
Indo-China: Laos(N) [29]; Vietnam(N) [29].
Description[29].

C. obcordata (Poiret)Bakh.f. [29]

Hedysarum obcordatum Poiret [29]; *Lourea obcordata* (Poiret)Desv. [29]; *Lourea reniforme* (Lour.)DC. [29].
Not climbing[29]; Herb[29]; Perennial[29].
Asia: China(N) [29]; India(N) [29]; Indonesia-ISO(N) [29]; Philippines(N) [29]. Indo-China: Cambodia(N) [29]; Laos(N) [29]; Thailand(N) [29]; Vietnam(N) [29].
Description[29]; Illustration[29].

var. **siamensia** (Craib)Ohashi [43]

Lourea paniculata var. *siamensia* Craib [32,43].
Not climbing[44]; Perennial[44].
Indo-China: Thailand(N) [32].
Description[32].

C. pierrei (Schindler)Ohashi [29]

C. translucida (Schindler)Ohashi [29]; *Lourea pierrei* Schindler [29]; *Lourea translucida* Schindler [29].
Not climbing[29]; Shrub[29]; Perennial[29].
Asia: Indonesia-ISO(N) [29]. Indo-China: Cambodia(N) [29]; Thailand(N) [43]; Vietnam(N) [29].
Description[29]; Illustration[29].

C. vespertilionis (L.f.)Bakh.f. [29]

Not climbing[29]; Herb[29]; Perennial[29].
Indo-China: Cambodia(N) [29]; Thailand(N) [29,32]; Vietnam(N) [29]. Asia: China(N) [29]; India(N) [29]; Indonesia-ISO(N) [29]; Malaysia-ISO(N) [29].
Description[29]; Illustration[29].
Medicine[29].

var. **vespertilionis** [29]

Hedysarum vespertilionis L.f. [29]; *Lourea vespertilionis* (L.f.)Desv. [29].
Indo-China: Cambodia(N) [29]; Thailand(N) [29,32]; Vietnam(N) [29]. Asia: China(N) [29]; India(N) [29]; Indonesia-ISO(N) [29]; Malaysia-ISO(N) [29].
Not climbing[29]; Herb[29]; Perennial[29].
Description[29]; Illustration[29].
Medicine[29].

var. **grandifolia** Dy Phon [29]

Not climbing[29]; Herb[29]; Perennial[29].
Indo-China: Vietnam(N) [29].
Description[29]; Illustration[29].

CODARIOCALYX Hassk.

Shrubs; two species in southeast Asia and tropical Australia.

C. gyroides (Link)Hassk. [31]

Desmodium bracteatum sensu M. Micheli [31]; *Desmodium gyroides* (Link)DC. [30,31]; *Desmodium oxalidifolium* Leveille,p.p. [31]; *Desmodium papuanum* C.White [31]; *Desmodium pseudogyroides* Miq. [31]; *Meibomia bracteata* (M.Micheli)Hoehne [31]; *Meibomia gyroides* (DC.)Kuntze [31].
Not climbing[31]; Shrub[31]; Perennial[31].
Indo-China: Cambodia(N) [30,31]; Thailand(N) [31]; Vietnam(N) [31]. Asia: Burma(N) [31]; India(N) [31]; Indonesia-ISO(N) [30]; Philippines(N) [30]; Sri Lanka(N) [31].
Description[30,31]; Distribution Map[31]; Illustration[31].

C. motorius (Houtt.)Ohashi [31]

C. gyrans (L.f.)Hassk. [31]; *Desmodium gyrans* (L.f.)DC. [30,31]; *Desmodium gyrans* var. *roylei* (Wight & Arn.)Bak. [31]; *Desmodium motorium* (Houtt.)Merr. [31]; *Meibomia gyrans* (L.f.)Kuntze [31].
Not climbing[31]; Shrub[31]; Perennial[31].
Indo-China: Cambodia(N) [30,31]; Laos(N) [30,31]; Thailand(N) [31]; Vietnam(N) [30,31]. Asia: Burma(N) [31]; China(N) [31]; India(N) [31]; Indonesia-ISO(N) [30]; Malaysia-ISO(N) [31]; Philippines(N) [30]; Sri Lanka(N) [31]; Taiwan(N) [31].
Description[30,31]; Distribution Map[31]]; Illustration[31].

DENDROLOBIUM (Wight & Arn.)Benth.

About 12 species, mainly in tropical Asia; shrubs or small trees.

D. baccatum (Schindler)Schindler [31]

Desmodium baccatum Schindler [31]; *Desmodium clovisii* Gagnep. [30,31].
Not climbing[31]; Shrub[31]; Perennial[31].
Indo-China: Thailand(N) [31]; Vietnam(N) [31].
Description[30,31]; Distribution Map[31]; Illustration[30,31].

D. lanceolatum (Dunn)Schindler [31]

Desmodium dunnii Merr. [31].
Not climbing[31]; Shrub or Tree[31]; Perennial[31].
Indo-China: Cambodia(N) [31]; Laos(N) [31]; Thailand(N) [31]; Vietnam(N) [31]. Asia: China(N) [31].
Description[30,31]; Distribution Map[31]; Illustration[31].

var. **lanceolatum** [31]

Desmodium lanceolatum sensu Gagnep. [30,31]; *Lespedeza lanceolata* Dunn [31].
Indo-China: Cambodia(N) [30]; Laos(N) [30]; Thailand(N) [31]; Vietnam(N) [31]. Asia: China(N) [31].
Not climbing[31]; Shrub or Tree[31]; Perennial[31].
Description[30,31]; Distribution Map[31]; Illustration[31].

var. **microcarpum** Ohashi [31]

Not climbing[31]; Shrub or Tree[31]; Perennial[31].
Indo-China: Thailand(N) [31].
Description[31]; Illustration[31].

D. olivaceum (Prain)Schindler [31]

Desmodium olivaceum Prain [31].
Not climbing[31]; Shrub[31]; Perennial[31].
Asia: Burma(N) [31]. Indo-China: Thailand(N) [31].
Description[31]; Distribution Map[31]; Illustration[31].

D. rostratum (Schindler)Schindler [31]

Desmodium rostratum Schindler [30,31].
Not climbing[31]; Shrub[31]; Perennial[31].
Indo-China: Vietnam(N) [31].
Description[30,31]; Distribution Map[31]; Illustration[31].

D. rugosum (Prain)Schindler [31]

Not climbing[31]; Shrub[31]; Perennial[31].
Indo-China: Cambodia(N) [31]; Laos(N) [31]; Thailand(N) [31]; Vietnam(N) [31]. Asia: Burma(N) [31].
Description[31]; Distribution Map[31]; Illustration[31].

var. **rugosum** [31]

Desmodium rugosum Prain [30,31].
Not climbing[31]; Shrub[31]; Perennial[31].
Indo-China: Laos(N) [31]; Thailand(N) [31]. Asia: Burma(N) [31].
Description[30,31]; Distribution Map[31]; Illustration[31].

var. **moniliferum** Ohashi [31]

D. wallichii (Prain)Schindler [31]; *Desmodium umbellatum* sensu Bak., p.p. [31]; *Desmodium wallichii* sensu Prain [31].
Not climbing[31]; Shrub[31]; Perennial[31].
Indo-China: Cambodia(N) [31]; Thailand(N) [31]; Vietnam(N) [31]. Asia: Burma(N) [31].
Description[31]; Distribution Map[31]; Illustration[31].

D. thorelii (Gagnep.)Schindler [31]

Desmodium olivaceum sensu Schindler [31]; *Desmodium olivaceum* var. *thorelii* Schindler [31]; *Desmodium thorelii* Gagnep. [30,31].
Not climbing[31]; Shrub[31]; Perennial[31].
Indo-China: Laos(N) [31]; Thailand(N) [31].
Description[30,31]; Distribution Map[31]; Illustration[30,31].

D. triangulare (Retz.)Schindler [31]

Not climbing[31]; Shrub[31]; Perennial[31].
Indo-China: Cambodia(N) [30]; Laos(N) [31]; Thailand(N) [31]; Vietnam(N) [31]. Asia: Burma(N) [31]; China(N) [31]; India(N) [31]; Malaysia-ISO(N) [31]; Sri Lanka(N) [31].
Description[31]; Distribution Map[31]; Illustration[31].

subsp. **triangulare** [31]

D. cephalotes (Roxb.)Benth. [31].; *Desmodium cephalotes* (Roxb.)Wight & Arn. [30,31]; *Desmodium recurvatum* sensu Benth. [31]; *Desmodium triangulare* (Retz.)Merr. [31]; *Hedysarum triangulare* Retz. [31]; *Meibomia cephalotes* (Roxb.)Kuntze [31].
Not climbing[31]; Shrub[31]; Perennial[31].
Indo-China: Cambodia(N) [30]; Laos(N) [31]; Thailand(N) [31]; Vietnam(N) [31]. Asia: Burma(N) [31]; China(N) [31]; India(N) [31]; Malaysia-ISO(N) [31]; Sri Lanka(N) [31].
Description[30,31]; Distribution Map[31]; Illustration[31].
Two forms exist: forma *triangulare* occurs in Indo-China, forma *villosum* Ohashi occurs in Burma [31].

subsp. **cephalotoides** (Craib)Ohashi [31]

D. cephalotoides (Craib)Schindler [31]; *Desmodium cephalotoides* Craib [30,31].
Not climbing[31]; Shrub[31]; Perennial[31].
Indo-China: Thailand(N) [31]. Asia: Burma(N) [31]; India(N) [31].
Description[30,31]; Distribution Map[31]; Illustration[31].

D. umbellatum (L.)Benth. [31]

D. australe (Willd.)Benth. [31]; *D. cumingianum* Benth. [31]; *Desmodium cumingianum* (Benth.) Benth. [31]; *Desmodium umbellatum* (L.)DC. [30,31]; *Desmodium umbellatum* var. *costatum* Craib [[31]; *Hedysarum australe* Willd. [31]; *Hedysarum ellipticum* Miq. [31]; *Hedysarum umbellatum* L. [31]; *Meibomia umbellata* (L.)Kuntze [31]; *Ormocarpum umbellatum* (L.)Desv. [31].
Not climbing[31]; Shrub or Tree[31]; Perennial[31].
Indo-China: Thailand(N) [31]; Vietnam(N) [30]. Asia: Burma(N) [31]; China(N) [31]; India(N) [[31]; Malaysia-ISO(N) [31]; Sri Lanka(N) [31]; Taiwan(N) [31].
Description[30,31]; Illustration[31].
The native habitat of forma *hirsutum* is unknown; forma *umbellatum* occurs in Indo-China [31].

D. ursinum (Schindler)Schindler [31]

Desmodium ursinum Schindler [30,31].
Not climbing[31]; Shrub[31]; Perennial[31].
Indo-China: Vietnam(N) [31].
Description[30,31]; Distribution Map[31]; Illustration[31].

DESMODIUM Desv.

Perhaps 300 species, mainly erect herbs. Pantropical but with major centres of diversity in eastern Asia and tropical South and Central America. Several species have been widely introduced as fodder crops, and the burr-like fruit segments may well have led to other accidental introductions. These deliberate and accidental introductions have led to taxonomic confusion in some areas.

D. **adscendens** (Sw.)DC. [31]

D. oxalidifolium Miq. [31]; *D. thwaitesii* Bak. [31]; *D. trifoliastrum* Miq. [31]; *Meibomia adscendens* (Sw.)Kuntze,p.p. [31]; *Meibomia thwaitesii* (Bak.)Kuntze [31]; *Meibomia trifoliastra* (Miq.)Kuntze [31].
Climbing or not[31]; Herb or Shrub[31]; Perennial[31].
Indo-China: Thailand(N) [31]. Asia: India(N) [31]; Malaysia-ISO(N) [31]; Sri Lanka(N) [31].
Description[31]; Illustration[31].

D. **amethystinum** Dunn [33]

Not climbing[33]; Shrub[33]; Perennial[33].
Indo-China: Thailand(N) [33]. Asia: China(N) [33].
Description[33]; Illustration[33].

D. **auricomum** Benth. [31].

Not climbing[31]; Herb[31]; Annual[31].
Indo-China: Cambodia(N) [30]; Thailand(N) [31]; Vietnam(N) [30]. Asia: Burma(N) [31]; Indonesia-ISO(N) [31].
Description[30,31]; Distribution Map[31]; Illustration[31].

D. **caudatum** (Thunb.)DC. [31]

Catenaria caudata (Thunb.)Schindler [31]; *Catenaria laburnifolia* (Poir.)Benth. [31]; *D. laburnifolium* (Poir.)DC. [31]; *Meibomia caudata* (Thunb.)Kuntze [31]; *Meibomia laburnifolia* (Poir.)Kuntze [31].
Not climbing[31]; Shrub[31]; Perennial[31].
Indo-China: Vietnam(N) [30]. Asia: Burma(N) [31]; China(N) [31]; India(N) [31]; Indonesia-ISO(N) [31]; Japan(N) [31]; Malaysia-ISO(N) [31]; Sri Lanka(N) [31]; Taiwan(N) [31].
Description[30,31]; Distribution Map[31]; Illustration[31].

D. **craibii** Ohashi [152]

Murtonia kerrii Craib [152].
Climbing[152]; Herb or Shrub[152]; Perennial[152].
Indo-China: Thailand(N) [152].
Description[152]; Illustration[152].
Very distinct; gievn subgeneric rank by Ohashi (Ref. 152).

D. **diffusum** DC. [153]

D. laxiflorum sensu auct. [153]; *D. recurvatum* (Roxb.)Wight & Arn. [153]; *D. unibotryosum* C.Chen & X.J.Cui [153]; *Hedysarum recurvatum* Roxb. [153].
Not climbing[153]; Shrub[153]; Perennial[153].
Indo-China: Vietnam(N) [153]. Asia: China(N) [153]; India(N) [153]; Taiwan(N) [153].
Description[35, 153].

D. **flexuosum** Benth. [31]

Meibomia flexuosa (Benth.)Kuntze [31].
Not climbing[31]; Shrub[31]; Perennial[31].
Indo-China: Thailand(N) [31]. Asia: Burma(N) [31].
Description[30,31]; Distribution Map[31]; Illustration[31].

D. **gangeticum** (L.)DC. [31]

D. polygonoides Bak. [31]; *Hedysarum gangeticum* L. [31]; *Hedysarum pseudogangeticum* Miq. [31]; *Meibomia gangetica* (L.)Kuntze [31]; *Meibomia polygonoides* (Bak.)Kuntze [31].
Climbing or not[31]; Shrub[31]; Perennial[31].
Indo-China: Cambodia(N) [31]; Laos(N) [31]; Thailand(N) [31]; Vietnam(N) [31]. Asia: Burma(N) [31]; China(N) [31]; India(N) [31]; Malaysia-ISO(N) [31]; Sri Lanka(N) [31]; Taiwan(N) [31].
Description[30,31]; Distribution Map[31]; Illustration[31].

D. griffithianum Benth. [31]

D. oxalidifolium Leveille [31]; *Meibomia griffithiana* (Benth.)Kuntze [31].
Climbing or not[31]; Herb or Shrub[31]; Perennial[31].
Indo-China: Laos(N) [31]; Thailand(N) [31]; Vietnam(N) [31]. Asia: Burma(N) [31]; China(N) [31]; India(N) [31].
Description[30,31]; Distribution Map[31]; Illustration[31].

D. harmsii Schindler [31]

Not climbing[31]; Shrub[31]; Perennial[31].
Indo-China: Vietnam(N) [31].
Description[30,31]; Distribution Map[31]; Illustration[31].
Endemic to S. Indo-China [31].

D. hayatae Ohashi [33]

Not climbing[33]; Shrub[33]; Perennial[33].
Indo-China: Thailand(N) [33].
Description[33]; Illustration[33].
Endemic to Thailand [33].

D. heterocarpon (L.)DC. [31]

Not climbing[31]; Herb or Shrub[31]; Perennial[31].
Indo-China: Cambodia(N) [31]; Laos(N) [31]; Thailand(N) [31]; Vietnam(N) [31]. Asia: Burma(N) [31]; China(N) [31]; India(N) [31]; Japan(N) [31]; Malaysia-ISO(N) [31]; Sri Lanka(N) [31]; Taiwan(N) [31].
Description[31]; Illustration[31].

subsp. **heterocarpon** [31]

D. buergeri Miq. [31]; *D. polycarpum* (Poir.)DC. [31]; *D. toppinii* Schindler [154].
Climbing or not[31]; Herb or Shrub[31]; Perennial[31].
Indo-China: Cambodia(N) [30]; Laos(N) [31]; Thailand(N) [31]; Vietnam(N) [31]. Asia: Burma(N) [31]; China(N) [31]; India(N) [31]; Japan(N) [31]; Malaysia-ISO(N) [31]; Sri Lanka(N) [31]; Taiwan(N) [31].
Description[31]; Illustration[31].
Three varieties of subsp. *heterocarpon* occur in Indo-China. Var. *gymnocarpum* Schindler occurs in Sri Lanka & India [31].

subsp. **angustifolium** Ohashi [154]

D. polycarpum var. *angustifolium* Craib [31]; *D. reticulatum* Benth. [31]; *D. reticulatum* var. *pilosum* Craib [31]; *Meibomia reticulata* (Benth.)Kuntze [31].
Climbing or not[31]; Herb or Shrub[31]; Perennial[31].
Indo-China: Cambodia(N) [30]; Laos(N) [30]; Thailand(N) [31]; Vietnam(N) [31]. Asia: Burma(N) [31]; China(N) [31].
Description[30,31]; Illustration[31].
Craib's varietal name is a *nomen nudum* [154].
Craib's var. *pilosum* is reduced to a forma by Ohashi [154].

var. **birmanicum** (Prain)Ohashi [31]

D. birmanicum Prain [31]; *D. oblongum* Kurz,p.p. [31].
Climbing or not[31]; Herb or Shrub[31]; Perennial[31].
Indo-China: Thailand(N) [31]. Asia: Burma(N) [31].
Description[31]; Illustration[31].

subsp. **ovalifolium** (Prain)Ohashi [154]

D. ovalifolium Merr. [154]; *D.polycarpum* DC. var. *ovalifolium* Prain [154].
Not climbing[154]; Herb[154]; Perennial[154].
Indo-China: Cambodia(N) [154]; Laos(N) [154]; Thailand(N) [154]; Vietnam(N) [154]. Asia: Indonesia-ISO(N) [154]; Malaysia-ISO(N) [154]; Philippines(N) [154].
Description[154]; Illustration[154].
The name *D. ovalifolium* has been attributed to Prain and to Gagnepain. Roxburgh's invalid name was in fact first validated by Merrill [154].

var. **strigosum** Van Meeuwen [31]

D. polycarpum (Poir.)DC.,p.p. [31]; *D. siliquosum* (Burm.f.)DC. [31].
Climbing or not[31]; Herb or Shrub[31]; Perennial[31].
Indo-China:Thailand(N) [31]; Vietnam(N) [31]. Asia: Burma(N) [31]; China(N) [31]; India(N) [31]; Malaysia-ISO(N) [31].
Description[31]; Illustration[31].

D. heterophyllum (Willd.)DC. [31]

D. triflorum var. *majus* Wight & Arn. [31]; *Hedysarum heterophyllum* Willd.; *Meibomia heterophylla* (Willd.)Kuntze [31].
Climbing or not[31]; Shrub[31]; Perennial[31].
Indo-China: Thailand(N) [31]; Vietnam(N) [4]. Asia: Burma(N) [31]; China(N) [31]; India(N) [31]; Malaysia-ISO(N) [31]; Borneo[31]; Sri Lanka(N) [31]; Taiwan(N) [31].
Description[31]; Illustration[31].
W.T. Tsang 29215: 'Vietnam' [4].

D. kingianum Prain [33]

Not climbing[33]; Shrub[33]; Perennial[33].
Indo-China: Cambodia(N) [33]; Laos(N) [33]; Thailand(N) [33]. Asia: Burma(N) [33].
Description[33]; Illustration[33].

D. laxiflorum DC. [31]

D. incanum sensu auctt. [31]; *D. scabrellum* Miq. [31]; *Meibomia laxiflora* (DC.)Kuntze [31].
Not climbing[31]; Shrub[31]; Perennial[31].
Indo-China: Laos(N) [30]; Thailand(N) [31]; Vietnam(N) [30]. Asia: India(N) [31]; Malaysia-ISO(N) [31]; Philippines(N) [154]; Taiwan(N) [31].
Description[30,31]; Distribution Map[31]; Illustration[31]].
Subsp. *lacei* is endemic to Burma [31].

D. laxum DC. [31]

Climbing or not[31]; Herb[31]; Perennial[31].
Indo-China: Cambodia(N) [31]; Laos(N) [30]; Thailand(N) [31]; Vietnam(N) [31]. Asia: China(N) [31]; India(N) [31]; Japan(N) [31]; Malaysia-ISO(N) [31]; Philippines(N) [31]; Taiwan(N) [31].
Description[31]; Distribution Map[31]; Illustration[31].

subsp. **laxum** [31]

D. austro-japonense Ohwi [31]; *D. bambusetorum* Miq. [31]; *D. gardneri* sensu auctt.,p.p. [31]; *D. laxiflorum* sensu Miq.,p.p. [31]; *D. laxum* var. *kiusianum* Matsum. [31]; *D. podocarpum* sensu auctt. [31]; *D. podocarpum* Miq.,p.p. [31]; *D. podocarpum* var. *gardneri* sensu Beddome [31]; *D. podocarpum* var. *indicum* Maxim.,p.p. [31]; *D. podocarpum* var. *laxum* (DC.)Bak. [31]; *Meibomia bambusetorum* (Miq.)Kuntze [31]; *Meibomia gardneri* Kuntze [31].
Climbing or not[31]; Herb[31]; Perennial[31].
Indo-China: Laos(N) [30]; Thailand(N) [31]; Vietnam(N) [31]. Asia: China(N) [31]; India(N) [31]; Japan(N) [31].
Description[30,31]; Distribution Map[31]; Illustration[31].

subsp. **laterale** (Schindler)Ohashi [31]

D. hainanense Isely [31]; *D. laterale* Schindler [31]; *D. laxum* sensu Matsum.,p.p. [31].
Climbing or not[31]; Herb[31]; Perennial[31].
Asia: China(N) [31]; Japan(N) [31]; Taiwan(N) [31].
Description[31]; Distribution Map[31]; Illustration[31].
This subspecies does not occur in Indo-China [31].

subsp. **leptopus** (Benth.)Ohashi [31]

D. gardneri Benth. [31]; *D. laxum* sensu auctt. [31]; *D. leptopus* Benth. [31]]; *D. tashiroi* Matsum. [31]; *Meibomia leptopus* (Benth.)Kuntze [31].
Not climbing[31]; Herb[31]; Perennial[31].
Indo-China: Cambodia(N) [31]; Laos(N) [31]; Thailand(N) [31]; Vietnam(N) [31]. Asia: China(N) [31]; Japan(N) [31]; Malaysia-ISO(N) [31]; Philippines(N) [31]; Taiwan(N) [31].
Description[31]; Distribution Map[31]; Illustration[31].

D. megaphyllum Zoll. [33]

Not climbing[33]; Shrub[33]; Perennial[33].
Indo-China: Thailand(N) [33]. Asia: Burma(N) [33]; China(N) [33]; India(N) [33]; Indonesia-ISO(N) [33]; Malaysia-ISO(N) [33].
Description[33]; Illustration[33].
Var. *glabrescens* does not occur in Indo-China. Found in Burma [33].

var. **megaphyllum** [33]

D. karensium Kurz [33]; *D. prainii* Schindler [33]; *D. rubescens* Miq. [33]; *D. scandens* Miq. [33]; *Meibomia megaphylla* (Zoll.)Kuntze [33].
Not climbing[33]; Shrub[33]; Perennial[33].
Indo-China: Thailand(N) [33]. Asia: Burma(N) [33]; China(N) [33]; India(N) [33]; Indonesia-ISO(N) [33]; Malaysia-ISO(N) [33].
Description[33]; Illustration[33].

D. microphyllum (Thunb.)DC. [31]

Not climbing[31]; Shrub[31]; Perennial[31].
Indo-China: Cambodia(N) [30]; Laos(N) [31]; Thailand(N) [31]; Vietnam(N) [31]. Asia: Burma(N) [31]; China(N) [31]; India(N) [31]; Japan(N) [31]; Malaysia-ISO(N) [31]; Sri Lanka(N) [31]; Taiwan(N) [31].
Description[30,31]; Illustration[31].
Two varieties occur. Only var. *microphyllum* is found in Indo-China. Var. *macrocarpum* Schindler is found in E. Nepal and Assam [31].

var. **microphyllum** [31]

D. parvifolium DC. [31]; *Meibomia microphylla* (Thunb.)Kuntze [31].
Not climbing[31]; Shrub[31]; Perennial[31].
Indo-China: Cambodia(N) [30]; Laos(N) [31]; Thailand(N) [31]; Vietnam(N) [31]. Asia: Burma(N) [31]; China(N) [31]; India(N) [31]; Japan(N) [31]; Malaysia-ISO(N) [31]; Sri Lanka(N) [31]; Taiwan(N) [31].
Description[31]; Illustration[31].

D. multiflorum DC. [33]

D. floribundum G.Don [33]; *D. mairei* Pampan. [33]; *D. nepalense* Ohashi [33]; *D. sambuense* (D.Don)DC. [33]; *Meibomia floribunda* (D.Don)Kuntze [33].
Not climbing[33]; Shrub[33]; Perennial[33].
Indo-China: Laos(N) [33]; Thailand(N) [33]. Asia: Burma(N) [33]; China(N) [33]; India(N) [33].
Description[33]; Illustration[33].

D. oblongum Benth. [31]

Meibomia oblonga (Benth.)Kuntze [31].
Not climbing[31]; Shrub[31]; Perennial[31].
Indo-China: Cambodia(N) [31]; Laos(N) [31]; Thailand(N) [31]; Vietnam(N) [31]. Asia: Burma(N) [31]; China(N) [31]; India(N) [31].
Description[30,31]; Distribution Map[31]; Illustration[31].

D. renifolium (L.)Schindler [31]

Not climbing[31]; Shrub[31]; Perennial[31].
Indo-China: Laos(N) [31]; Thailand(N) [31]. Asia: Burma(N) [31]; China(N) [31]; India(N) [31]; Malaysia-ISO(N) [31]; Taiwan(N) [31].
Description[30,31]; Illustration[31].

var. **renifolium** [31]

D. reniforme (L.)DC. [30,31]; *Meibomia reniformis* (L.)Kuntz [31].
Not climbing[31]; Shrub[31]; Perennial[31].
Indo-China: Laos(N) [30]; Thailand(N) [31]. Asia: Burma(N) [31]; China(N) [31]; India(N) [31]; Malaysia-ISO(N) [31]; Taiwan(N) [31].
Description[30,31]; Illustration[31].

var. **oblatum** (Kurz)Ohashi [31]

D. oblatum Kurz [30,31]; *Meibomia oblata* (Kurz)Kuntze [31].
Not climbing[31]; Shrub[31]; Perennial[31].
Indo-China: Laos(N) [31]; Thailand(N) [31]. Asia: Burma(N) [31]; China(N) [31]; India(N) [31]; Malaysia-ISO(N) [31]; Taiwan(N) [31].
Description[30,31]; Illustration[31].

D. repandum (Vahl)DC. [31]

Anarthrosyne scalpe Klotzsch [31]; *D. caffrum* Eckl. & Zeyh. [31]; *D. scalpe* DC. [30,31]; *D. strangulatum* Wight & Arn. [31]; *Hedysarum ignescens* Miq. [31]; *Meibomia repanda* (Vahl)Kuntze [31]; *Meibomia scalpe* (DC.)Kuntze [31].
Not climbing[31]; Herb[31]; Perennial[31].
Indo-China: Laos(N) [31]; Thailand(N) [31]; Vietnam(N) [31]. Asia: Burma(N) [31]; China(N) [31]; India(N) [31]; Malaysia-ISO(N) [31]; Borneo[31]; Sri Lanka(N) [31].
Description[30,31]; Illustration[31].

D. rubrum (Lour.)DC. [31]

Climbing or not[31]; Shrub[31]; Perennial[31].
Indo-China: Vietnam(N) [31]. Asia: China(N) [31].
Description[31]; Distribution Map[31]; Illustration[31].

var. **rubrum** [31]

D. carlesii Schindler [31].
Climbing or not[31]; Shrub[31]; Perennial[31].
Indo-China: Vietnam(N) [31]. Asia: China(N) [31].
Description[31]; Distribution Map[31]; Illustration[31].

var. **macrocarpum** Ohashi [31]

Climbing or not[31]; Shrub[31]; Perennial[31].
Indo-China: Vietnam(N) [31].
Description[31]; Distribution Map[31]; Illustration[31].

D. schubertiae Ohashi [34]

Not climbing[34]; Shrub[34]; Perennial[34].
Indo-China: Cambodia(N) [34]; Vietnam(N) [34].
Description[34]; Illustration[34].

D. sequax Wall. [33]

D. dasylobum Miq. [33]; *D. hamulatum* Franch. [33]; *D. sequax* var. *sinuatum* (Miq.)Hosokawa [33]; *D. sinuatum* Bak. [33]; *D. strangulatum* var. *sinuatum* Miq. [33]; *Dollinera sequax* Hochr. [33]; *Meibomia dasyloba* (Miq.)Kuntze [33]; *Meibomia sequax* (Wall.)Kuntze [33]; *Meibomia sinuata* (Miq.)Kuntze [33].
Not climbing[33]; Shrub[33]; Perennial[33].
Indo-China: Vietnam(N) [30]. Asia: Burma(N) [33]; China(N) [33]; India(N) [33]; Indonesia-ISO(N) [33]; Malaysia-ISO(N) [33]; Taiwan(N) [33].
Description[33]; Illustration[33].

D. siamense (Schindler)Craib [33]

Not climbing[33]; Shrub[33]; Perennial[33].
Indo-China: Thailand(N) [33].
Description[33]; Illustration[33].
Endemic to Thailand [33].

D. strigillosum Schindler [31]

Climbing[31]; Shrub[31]; Perennial[31].
Indo-China: Cambodia(N) [31]; Laos(N) [31]; Vietnam(N) [31]. Asia: Burma(N) [31].
Description[31]; Illustration[31].

D. styracifolium (Osbeck)Merr. [31]

D. capitatum (Burm.f.)DC. [31]; *D. celebicum* Schindler [31]; *D. retroflexum* (L.)DC. [30,31]; *Meibomia capitata* (Burm.f.)Kuntze [31]; *Uraria retroflexa* (L.)Drake [31].
Climbing or not[31]; Shrub[31]; Perennial[31].
Indo-China: Cambodia(N) [30]; Thailand(N) [31]; Vietnam(N) [30]. Asia: Burma(N) [31]; China(N) [31]; India(N) [31]; Malaysia-ISO(N) [31]; Sri Lanka(N) [31].
Description[30,31]; Illustration[31].
Toxins[30].
Dry plant gives off a strong odour of coumarin [30].

D. teres Benth. [31]

Meibomia teres (Benth.)Kuntze [31].
Not climbing[31]; Shrub[31]; Perennial[31].
Indo-China: Laos(N) [31]; Thailand(N) [31]. Asia: Burma(N) [31].
Description[31]; Distribution Map[31]; Illustration[31].
Probably more widely distributed in Indo-China [31].

D. triflorum (L.)DC. [31]

Meibomia triflora (L.)Kuntze [31].
Not climbing[31]; Herb or Shrub[31]; Perennial[31].
Indo-China: Thailand(N) [31]; Vietnam(N) [30]. Asia: Burma(N) [31]; China(N) [31]; India(N) [31]; Malaysia-ISO(N) [31]; Sri Lanka(N) [31]; Taiwan(N) [31].
Description[30,31]; Illustration[31].

D. velutinum (Willd.)DC. [31]

Not climbing[31]; Shrub[31]; Perennial[31].
Indo-China: Cambodia(N) [31]; Laos(N) [31]; Thailand(N) [31]; Vietnam(N) [31]. Asia: Burma(N) [31]; China(N) [31]; India(N) [31]; Malaysia-ISO(N) [31]; Sri Lanka(N) [31]; Taiwan(N) [31].
Description[31]; Illustration[31].

subsp. **velutinum** [31]

Anarthrosyne cordata Klotzsch [31]; *D. lasiocarpum* (P.Beauv.)DC. [31]; *D. latifolium* (Ker)DC. [31]; *D. plukenetii* (Wight. & Arn.)Merr. & Chun [31]; *D. virgatum* Prain [31]; *Meibomia velutina* (Willd.)Kuntze [31]; *Pseudarthria cordata* (Klotzsch)C.Muell. [31].
Not climbing[31]; Shrub[31]; Perennial[31].
Indo-China: Cambodia(N) [30]; Laos(N) [31]; Thailand(N) [31]; Vietnam(N) [30]. Asia: Burma(N) [31]; China(N) [31]; India(N) [31]; Malaysia-ISO(N) [31]; Sri Lanka(N) [31]; Taiwan(N) [31].
Description[30,31]; Illustration[31].
Two varieties of the subspecies exist. Var. *sikkimense* (Schindler)Ohashi occurs in Nepal & Sikkim. Var. *velutinum* is found in Indo-China [31].

subsp. **longibracteatum** (Schindler)Ohashi [31]

D. longibracteatum Schindler [31]; *D. rufihirsutum* Craib [31]; *D. velutinum* var. *longibracteatum* (Schindler)Van Meeuwen [31].
Not climbing[31]; Shrub[31]; Perennial[31].
Asia: Burma(N) [31]; China(N) [31]; India(N) [31]; Indonesia-ISO(N) [31]. Indo-China: Laos(N) [31]; Thailand(N) [31].
Description[31]; Illustration[31].

D. vidalii Ohashi [156]

Not climbing[156]; Herb[156]; Perennial[156].
Indo-China: Laos(N) [156].
Description[156]; Illustration[156].

D. zonatum Miq. [31]

D. ormocarpoides sensu auctt. [31]; *D. shimadai* Hayata [31]; *Meibomia zonata* (Miq.)Kuntze [31].
Climbing or not[31]; Shrub[31]; Perennial[31].
Indo-China: Laos(N) [31]; Thailand(N) [31]; Vietnam(N) [30]. Asia: Burma(N) [31]; China(N) [31]; India(N) [31]; Malaysia-ISO(N) [31]; Sri Lanka(N) [31]; Taiwan(N) [31].
Description[30,31]; Distribution Map[31]; Illustration[30,31].

DICERMA DC.

A single shrubby species; tropical Asia to Australia.

D. biarticulatum (L.)DC. [31]

Climbing or not[31]; Shrub[31]; Perennial[31].
Indo-China: Cambodia(N) [30]; Laos(N) [30]; Thailand(N) [31]; Vietnam(N) [30]. Asia: Burma(N) [31]; China(N) [31]; India(N) [31]; Indonesia-ISO(N) [31]; Malaysia-ISO(N) [31]; Sri Lanka(N) [31].
Description[31]; Distribution Map[31]; Illustration[31].
Other subspecies in Australasia and Burma [31].

subsp. **biarticulatum** [31]

D. biarticulatum var. *collettii* Schindler [31]; *Aphyllodium biarticulatum* Gagnep. [30,31]; *Desmodium biarticulatum* (L.)F.v.Muell. [31]; *Meibomia biarticulata* (L.)Kuntze [31].
Climbing or not[31]; Shrub[31]; Perennial[31].
Indo-China: Cambodia(N) [30]; Laos(N) [30]; Thailand(N) [31]; Vietnam(N) [30]. Asia: Burma(N) [31]; China(N) [31]; India(N) [31]; Indonesia-ISO(N) [31]; Malaysia-ISO(N) [31]; Sri Lanka(N) [31].
Description[30,31]; Distribution Map[31]; Illustration[31].

DROOGMANSIA De Wild.

A small and mainly African genus of perhaps 20 species.

D. godefroyana (Kuntze)Schindler [29]

Meibomia godefroyana Kuntze [29]; *Tadehagi godefroyanum* (Kuntze)Ohashi [31].
Not climbing[29]; Shrub[29]; Perennial[29].
Indo-China: Cambodia(N) [29]; Laos(N) [29]; Thailand(N) [32]; Vietnam(N) [29].
Description[29]; Distribution Map[31]; Illustration[29,31].

HEGNERA Schindler

Often included within *Desmodium* but kept separate by Ohashi.

H. obcordata (Miq.)Schindler [31]

Desmodium obcordatum (Miq.)Kurz [30,31]; *Meibomia obcordata* (Miq.)Kuntze [31]; *Uraria obcordata* Miq. [31].
Climbing[31]; Shrub[31]; Perennial[31].
Indo-China: Cambodia(N) [31]; Laos(N) [31]; Thailand(N) [31]; Vietnam(N) [31]. Asia: Burma(N) [31]; Indonesia-ISO(N) [31]; Malaysia-ISO(N) [31].
Description[30,31]; Distribution Map[31]; Illustration[31].

KUMMEROWIA Schindler

Two species, in eastern Asia and North America.

K. striata (Thunberg)Schindler [29]

Lespedeza stipulacea Maxim. [29].
Indo-China: Laos(N) [30]; Vietnam(N) [30]. Asia: China(N) [30]; Japan(N) [30].
Not climbing[29]; Herb or Shrub[29]; Perennial[29].
Description[29,30]; Illustration[29,30].
Forage[30].

LESPEDEZA Michaux

Perennial herbs, sometimes woody; 30-40 species in temperate North America, eastern Asia and Australia. Some are cultivated as forage and green manure.

L. juncea (L.f.)Persoon [29]

L. aitchisonii Ricker [29].
Not climbing[37]; Herb or Shrub[37]]; Perennial[37].
Description[37,38]; Distribution Map[37,38].
Environmental[38]; Forage[38].

var. **sericea** (Thunb.)Lace & Hemsley [29,37]
L. cuneata (Dum.-Cours.)G.Don [29,38]; *L. juncea* subsp. *sericea* (Thunb.)Steenis [29]; *L. sericea* (Thunb.)Miq. [29].
Not climbing[29]; Shrub[29]; Perennial[29].
Indo-China: Laos(N) [29]; Thailand(N) [4]; Vietnam(N) [29]. Asia: Burma(N) [29]; China(N) [29]; India(N) [29]; Indonesia-ISO(N) [37]; Japan(N) [37]; Malaysia-ISO(N) [37]; Philippines(N) [37]; Taiwan(N) [29].
Description[29,37]; Distribution Map[37]; Illustration[29,38].
Environmental[38]; Forage[38].
The only representative of the genus occuring in both temperate & tropical climates [37]. *Floto* 7303: 'Thailand' [4].

MECOPUS Bennett

A single species in India and southeast Asia.

M. nidulans Bennett [29]
Not climbing[29]; Herb[29]; Perennial[29].
Indo-China: Cambodia(N) [29]; Laos(N) [29]; Thailand(N) [29]; Vietnam(N) [29]. Asia: China(N) [29]; India(N) [29]; Indonesia-ISO(N) [29].
Description[29]; Illustration[29].
Environmental[29]; Medicine[29].

PHYLACIUM Bennett

Two species in southeast Asia, extending through Malesia to Australia.

P. bracteosum Bennett [36]
Indo-China: Thailand(N) [36]. Asia: Indonesia-ISO(N) [36]; Malaysia-ISO(N) [36].
Description[36]; Distribution Map[36]; Illustration[36].

P. majus Collett & Hemsley [29]
Climbing[29]; Herb[29]; Perennial[29].
Indo-China: Laos(N) [29]; Thailand(N) [29]. Asia: Burma(N) [29].
Description[29,36]; Distribution Map[36]; Illustration[29,36].

PHYLLODIUM Desv.

Six species in southern Asia and Malesia, extending to Australia.

P. elegans (Lour.)Desv. [31]
Desmodium blandum Van Meeuwen [31]; *Desmodium elegans* (Lour.)Benth. [31,32]; *Meibomia elegans* (Lour.)Kuntze,p.p. [31].
Not climbing[31]; Shrub[31]; Perennial[31].
Indo-China: Cambodia(N) [31]; Laos(N) [31]; Thailand(N) [31]; Vietnam(N) [31]. Asia: China(N) [31]; Indonesia-ISO(N) [31].
Description[31]; Distribution Map[[31]; Illustration[31].

P. insigne (Prain)Schindler [31]
Desmodium insigne Prain [30,31].
Not climbing[31]; Shrub[31]; Perennial[31].
Indo-China: Thailand(N) [31]. Asia: Burma(N) [31].
Description[30,31]; Distribution Map[31]; Illustration[31].

P. kurzianum (Kuntze)Ohashi [31]
Desmodium grande Kurz [31]; *Desmodium kurzii* Craib [31]; *Meibomia kurziana* Kuntze [31]; *Phyllodium grande* (Kurz)Schindler [31]; *Phyllodium kurzii* (Craib)Chun [31].
Not climbing[31]; Shrub[31]; Perennial[31].
Indo-China: Thailand(N) [31]; Vietnam(N) [4]. Asia: Burma(N) [31]; China(N) [31].
Description[31]; Distribution Map[31]; Illustration[31].
W.T. Tsang 30677: 'Vietnam' [4].

P. longipes (Craib)Schindler [31]

Desmodium longipes Craib [30,31,32]; *Desmodium tonkinense* Schindler [[31].
Not climbing[31]; Shrub[31]; Perennial[31].
Indo-China: Cambodia(N) [31]; Laos(N) [31]; Thailand(N) [31]; Vietnam(N) [31]. Asia: Burma(N) [31]; China(N) [31].
Description[30,31]; Distribution Map[31]; Illustration[31].

P. pulchellum (L.)Desv. [31]

Desmodium pulchellum (L.)Benth. [30,31]; *Meibomia pulchella* (L.)Kuntze [31].
Not climbing[31]; Shrub[31]; Perennial[31].
Indo-China: Laos(N) [30]; Thailand(N) [31]; Vietnam(N) [31]. Asia: Burma(N) [31]; China(N) [31]; India(N) [31]; Indonesia-ISO(N) [31]; Malaysia-ISO(N) [31]; Sri Lanka(N) [31]; Taiwan(N) [[31].
Description[30,31]; Distribution Map[31]; Illustration[31].

P. vestitum Benth. [31]

Desmodium vestitum Bak. [30,31].
Not climbing[30]; Shrub[30]; Perennial[30].
Indo-China: Laos(N) [31]; Thailand(N) [31]; Vietnam(N) [31]. Asia: Burma(N) [31]; Indonesia-ISO(N) [30].
Description[30,31]; Distribution Map[31]; Illustration[30,31].

PYCNOSPORA Wight & Arn.

A single remarkably wide-ranging species, extending from Africa to southeast Asia, Malesia and tropical Australia. A small herb.

P. lutescens (Poiret)Schindler [29]

P. hedysaroides Wight & Arn. [29]; *P. nervosa* Wight & Arn. [29,30]; *Hedysarum lutescens* Poiret [29].
Not climbing[29]; Shrub[29]; Perennial[29].
Indo-China: Cambodia(N) [29]; Laos(N) [29]; Thailand(N) [32]; Vietnam(N) [29]. Asia: China(N) [29]; India(N) [29]; Indonesia-ISO(N) [29]; Philippines(N) [29]; Taiwan(N) [29] Australasia: Australia(N) [29].
Description[29]; Illustration[29].

TADEHAGI Ohashi

Three shrubby species in eastern Asia, Malesia and Australia.

Tadehagi rodgeri (Schindler)Ohashi [157]

T. kerrii (Schindler)Ohashi [157]; *T. triquetrum* (L.)Ohashi subsp. *rodgeri* (Schindler)Ohashi [157]; *Desmodium kerrii* (Schindler)Craib [31]; *Pteroloma kerrii* Schindler [31]; *Pteroloma rodgeri* Schindler [157]; *Pteroloma triquetrum* (L.)Benth. subsp. *rodgeri* (Schindler)Ohashi [157].
Not climbing[31]; Shrub[31]; Perennial[31].
Indo-China: Laos(N) [157]; Thailand(N) [31]. Asia: Burma(N) [157].
Description[31]; Distribution Map[31]; Illustration[31].
Endemic to Thailand [31].

T. triquetrum (L.)Ohashi [31]

Climbing or not[31]; Shrub[31]; Perennial[31].
Indo-China: Laos(N) [31]; Thailand(N) [31]; Vietnam(N) [31]. Asia: Burma(N) [31]; China(N) [31]; India(N) [31]; Indonesia-ISO(N) [31]; Malaysia-ISO(N) [31]; Philippines(N) [31]; Sri Lanka(N) [31].
Description[31]; Distribution Map[31]; Illustration[31].
Five subspecies occur. Only the typical subspecies occurs in Indo-China [31].

subsp. **triquetrum** [31]

Desmodium triquetrum (L.)DC. [31]; *Desmodium triquetrum* subsp. *genuinum* Prain [31]; *Meibomia triquetra* (L.)Kuntze [31]; *Pteroloma triquetrum* (L.)Benth. [31].
Climbing or not[31]; Shrub[31]; Perennial[31].
Indo-China: Laos(N) [31]; Thailand(N) [31]; Vietnam(N) [31]. Asia: Burma(N) [31]; China(N) [31]; India(N) [31]; Indonesia-ISO(N) [31]; Malaysia-ISO(N) [31]; Sri Lanka(N) [31].
Description[31]; Distribution Map[31]; Illustration[31].

TRIFIDACANTHUS Merr.

One or two species of spiny shrubs, found in southeast Asia and Malesia.

T. unifoliolatus Merr. [29]

Desmodium horridum Steenis [29]; *Desmodium unifoliolatum* (Merr.)Steenis [158].
Not climbing[29]; Shrub[29]; Perennial[29].
Asia: China(N) [29]; Indonesia-ISO(N) [29]; Philippines(N) [29]. Indo-China: Vietnam(N) [29].
Description[29]; Illustration[29].

URARIA Desv.

About 20 species or herbs or small shrubs. Two species in Africa, the rest in southeast Asia, Malesia and Australia.

U. acaulis Schindler [29]

Not climbing[29]; Herb[29]; Perennial[29].
Indo-China: Cambodia(N) [29]; Laos(N) [29]; Thailand(N) [29]; Vietnam(N) [29].
Description[29].

U. acuminata Kurz [29]

Not climbing[29]; Shrub[29]; Perennial[29].
Asia: Burma(N) [29]. Indo-China: Laos(N) [29]; Thailand(N) [29]; Vietnam(N) [29].
Description[29,30].

U. balansae Schindler [29]

Not climbing[29]; Herb[29]; Perennial[29].
Indo-China: Vietnam(N) [29].
Description[29]; Illustration[29].
Endemic to northern Vietnam [29].

U. campanulata (Benth.)Gagnep. [29]

U. formosana Hayata [29]; *Christia campanulata* (Benth.)Thoth. [29]; *Lourea campanulata* Benth. [29].
Not climbing[29]; Herb or Shrub[29]; Perennial[29].
Indo-China: Thailand(N) [29]; Vietnam(N) [29]. Asia: Burma(N) [29]; India(N) [29].
Description[29]; Illustration[29].

U. cochinchinensis Schindler [29]

U. collettii sensu auct. [30].
Not climbing[29]; Shrub[29]; Perennial[29].
Indo-China: Cambodia(N) [29]; Laos(N) [29]; Vietnam(N) [29].
Description[29]; Illustration[29].
Medicine[29].

U. cordifolia Wall. [29]

U. latifolia Prain [29,30,32]; *Urariopsis cordifolia* Schindler [29,30].
Not climbing[29]]; Shrub[29]; Perennial[29].
Indo-China: Cambodia(N) [29]; Laos(N) [29]; Thailand(N) [29]; Vietnam(N) [29]. Asia: Burma(N) [29]; India(N) [29]; Indonesia-ISO(N) [29].
Description[29]; Distribution Map[45]; Illustration[30].
Domestic[29].

U. crinita DC. [29]

U. crinita var. *macrostachya* (Wall.)Schindler [29].
Not climbing[29]; Shrub[29]; Perennial[29].
Indo-China: Cambodia(N) [29]; Laos(N) [29]; Thailand(N) [29]; Vietnam(N) [29]. Asia: China(N) [29]; India(N) [29]; Indonesia-ISO(N) [29]; Malaysia-ISO(N) [29]; Philippines(N) [29]; Taiwan(N) [29].
Description[29].
Environmental[29]; Medicine[29].

U. lacei Craib [[29]

U. clarkei Gagnep. [29,30,32]; *U. pulchra* Haines [29].
Not climbing[29]; Shrub[29]; Perennial[29].
Indo-China: Laos(N) [29]; Thailand(N) [29]; Vietnam(N) [29]. Asia: Burma(N) [29]; China(N) [29].
Description[29,30].

U. lagopodioides DC. [29]

U. alopecuroides (Roxb.)Sweet [29]; *U. cylindracea* Benth. [29].
Not climbing[29]; Herb or Shrub[29]; Perennial[29].
Indo-China: Cambodia(N) [29]; Laos(N) [29]; Thailand(N) [29]; Vietnam(N) [29]. Asia: China(N) [29]; India(N) [29]; Taiwan(N) [29].
Description[29].
Medicine[29].

U. picta DC. [29]

Not climbing[29]; Shrub[29]; Perennial[29].
Indo-China: Cambodia(N) [29]; Thailand(N) [30]; Vietnam(N) [29]. Asia: China(N) [29]; India(N) [29]; Indonesia-ISO(N) [29]; Malaysia-ISO(N) [29]; Taiwan(N) [29].
Description[29].

U. pierrei Schindler [29]

Not climbing[29]; Herb or Shrub[29]; Perennial[29].
Indo-China: Cambodia(N) [29]; Thailand(N) [29].
Description[29,30]; Illustration[30].

U. poilanei Dy Phon [29]

Not climbing[29]; Shrub[29]; Perennial[29].
Indo-China: Laos(N) [29]; Thailand(N) [29].
Description[29]; Illustration[29].

U. rotundata Craib [4]

Not climbing[4]; Herb[4]; Perennial[4].
Indo-China: Thailand(N) [4].
Description[46].
A.F.G. Kerr 2136: 'Siam; creeping herb' [4].

U. rufescens (DC.)Schindler [29]

U. gracilis Prain [29]; *U. hamosa* Wight & Arn. [30]; *U. paniculata* Hassk. [29]; *Desmodium rufescens* DC. [29]; *Meibomia rufescens* Kuntze [29].
Not climbing[29]; Shrub[29]; Perennial[29].
Indo-China: Cambodia(N) [29]; Laos(N) [29]; Thailand(N) [29]; Vietnam(N) [29]. Asia: Burma(N) [29]; India(N) [29]; Indonesia-ISO(N) [29].
Description[29].
Medicine[29].

EUCHRESTEAE

EUCHRESTA Bennett

A small genus of about five species confined to east and southeast Asia.

E. horsfieldii (Leschen.)Bennett [29]

Andira horsfieldii Leschen. [29].
Climbing or not[29]; Shrub[29]; Perennial[29].
Indo-China: Laos(N)[29]; Thailand(N)[29]; Vietnam(N)[29]. Asia: China(N) [143,144]; India(N) [51]; Indonesia-ISO(N)[29].
Description[29,51], Illustration[29,51].
Medicine[29].

var. horsfieldii [29]

Climbing or not[29]; Shrub[29]; Perennial[29].
Indo-China: Thailand(N)[29]; Vietnam(N)[29]. Asia: China(N) [143,144]; India(N)[51]; Indonesia-ISO(N)[29].
Description[29,51], Illustration[29,51].
Medicine[29]

var. laotica Dy Phon [29]

Indo-China: Laos(N)[29].
Climbing or not[29]; Shrub[29]; Perennial[29].
Description[29], Illustration[29].
Variety endemic to Laos [29].

GALEGEAE

ASTRAGALUS L.

Probably the largest genus of flowering plants, with in the region of 3000 species. Most are annual or perennial herbs or sub-shrubs. They are most diverse in western and central Asia, but occur throughout the north temperate zone, extending into the tropics only on mountains.

A. sinicus L. [29]

A. lotoides Lam. [29]
Climbing or not[29]; Herb[29]; Annual[29].
Indo-China: Vietnam(N)[29]. Asia: China(N)[29]; Japan(N)[29]; Taiwan(N)[29].
Description[29]; Illustration[29].

GUELDENSTAEDTIA Fischer

Mostly small perennial herbs of montane habitats, occurring mainly in the Himalayan region but extending into Siberia. About 12 species.

G. verna (Georgi)Boriss. [29]

G. delavayi Franchet [29]; *G. mirpourensis* Bak. [29]; *G. pauciflora* (Pallas)Fischer [29]; *Amblyotropis multiflora* (Bunge)Kitagawa [29]; *Astragalus pauciflorus* Pallas [29]; *Astragalus vernus* Georgi [29].
Climbing or not[29]; Herb[29]; Perennial[29].
Indo-China: Laos(N)[29]. Asia: Burma(N)[29]; China(N)[29].
Description[29]; Illustration[29].

INDIGOFEREAE

CYAMOPSIS DC.

A genus of three species, all herbs, native to the drier areas of Africa, Arabia and Pakistan. One is widely cultivated as a vegetable and for the seed gum (Guar Gum).

C. tetragonoloba (L.)Taub. [29]

C. psoralioides (Lam.)DC. [29]; *Psoralea tetragonoloba* L. [29].
Not climbing [29]; Herb[29]; Perennial[29].
Indo-China: Vietnam(I) [29]. Asia: India(N) [104].
Description[29]; Illustration[29].
Forage[29].
Cultivated and naturalized [29].

INDIGOFERA L.

A pantropical genus of about 700 species, mostly herbs, but some shrubs. Most diverse in Africa. Some are cultivated as dye-plants. Recently some of the species (not in southeast Asia) have been segregated as *Microcharis* and *Indigastrum*.

I. aralensis Gagnep. [29]

Not climbing [29]; Shrub[29]; Perennial[29].
Indo-China: Cambodia(N) [29]; Vietnam(N) [29].
Description[29,52].

I. arrecta A.Rich. [29]

Not climbing [29]; Shrub[29]; Perennial[29].
Indo-China: Laos(I) [29]; Vietnam(I) [29].
Description[29,52]; Illustration[52].
Chemical Products[52]; Environmental[52].
Java Indigo[52].
Native to tropical Africa, Indian Ocean islands and Middle East. It is often cultivated in Indo-China, or is an escape from cultivation [29].

I. atropurpurea Hornem. [52]

Not climbing [52]; Shrub[52]; Perennial[52].
Indo-China: Vietnam(N) [52]. Asia: Bangladesh(N) [52]; Burma(N) [52]; China(N) [52]; India(N) [52]; Nepal(N) [52]; Sri Lanka(N) [52].
Description[52]; Illustration[52].

I. banii Dang Khoi & Yakovlev [29]

Not climbing [29]; Herb[29]; Perennial[29].
Indo-China: Vietnam(N) [29].
Description[29,95]; Distribution Map[95]; Illustration[95].

I. caloneura Kurz [29]

I. oblonga Craib [29].
Not climbing [29]; Shrub[29]; Perennial[29].
Indo-China: Laos(N) [29]; Thailand(N) [29]; Vietnam(N) [29]. Asia: Burma(N) [29].
Description[29,52]; Illustration[29,52].
Recorded from India by Nguyen van Thuan et al. (1987; ref.29) but not by Sanjappa (1992; ref. 104) or by De Kort & Thijsse (1984; ref. 520.

I. cassioides DC. [29]

I. elliptica Roxb. [29]; *I. pulchella* Roxb. [29].
Not climbing [29]; Shrub[29]; Perennial[29].
Indo-China: Laos(N) [29]; Thailand(N) [29]; Vietnam(N) [29]. Asia: Burma(N) [29]; China(N) [29]; India(N) [29]; Nepal(N) [104]; Pakistan(N) [29].
Description[29]; Illustration[52].
Food or Drink[52]; Medicine[52].

I. caudata Dunn [29]

Not climbing [29]; Shrub[29]; Perennial[29].
Indo-China: Laos(N) [29]. Asia: China(N) [29].
Description[29,52]; Illustration[52].
Medicine[29].

I. colutea (Burm.f.)Merr. [29]

I. viscosa Lam. [29]; *Galega colutea* Burm.f. [29].
Not climbing [29]; Herb[29]; Annual or perennial[29].
Indo-China: Thailand(N) [52]; Vietnam(N) [29]. Asia: Burma(N) [29]; India(N) [29]; Indonesia-ISO(N) [29]; Sri Lanka(N) [29].
Description[29,52]; Illustration[52].
Widespread throughout the Old World (Schrire pers.comm.).

I. dosua D.Don [29]

Not climbing [29]; Shrub[29]; Perennial[29].
Indo-China: Laos(N) [29]; Thailand(N) [29]; Vietnam(N) [29]. Asia: Burma(N) [29]; China(N) [29]; India(N) [29]; Indonesia-ISO(N) [29].
Description[29,52]; Illustration[52].
Food or Drink[52]; Forage[52]; Medicine[52].

I. emmae De Kort & Thijsse [52]

Not climbing [52]; Herb or Shrub[52]; Perennial[52].
Indo-China: Thailand(N) [52]. Asia: Burma(N) [52]; India(N) [52].
Description[52].

I. galegoides DC. [29]

I. finlaysoniana Ridley [52]; *I. uncinata* Roxb. [29].
Not climbing [29]; Shrub[29]; Perennial[29].
Indo-China: Cambodia(N) [29]; Laos(N) [29]; Thailand(N) [29]; Vietnam(N) [29]. Asia: Burma(N) [29]; China(N) [29]; India(N) [29]; Indonesia-ISO(N) [29]; Malaysia-ISO(N) [29]; Sri Lanka(N) [29]; Taiwan(N) [29].
Description[29,52]; Distribution Map[52]; Illustration[52].
Toxins[52].

I. glabra L. [29]

I. pentaphylla Murray [29].
Not climbing [29]; Herb[29]; Annual[29].
Indo-China: Thailand(N) [52]; Vietnam(N) [29]. Asia: Burma(N) [104]; India(N) [29]; Sri Lanka(N) [29].
Description[29]; Illustration[52].

I. hendecaphylla Jacq. [52,54]

Not climbing [52]; Herb or Shrub[52]; Perennial[52].
Indo-China: Cambodia(N) [52]; Thailand(N) [52]; Vietnam(N) [52]. Asia: Burma(N) [52]; India(N) [52]; Indonesia-ISO(N) [52]; Malaysia-ISO(N) [52]; Philippines(N) [52]; Sri Lanka(N) [52].
Description[52,54].
Environmental[52]; Forage[52]; Toxins[52].

var. **hendecaphylla** [52,54]

I. bolusii N.E. Br. [54]; *I. endecaphylla* sensu Poir. [54]; *I. hendecaphylla* var. *angustata* Harv. [54]; *I. onobrychioides* Baillon [54]; *I. pectinata* Bak. [54]; *I. spicata* sensu auct., non Forsk. [54].
Not climbing [52]; Herb or Shrub[52]; Perennial[52].
Indo-China: Cambodia(N) [52]; Thailand(N) [52]; Vietnam(N) [52]. Asia: Burma(N) [52]; India(N) [52]; Indonesia-ISO(N) [52]; Malaysia-ISO(N) [52]; Philippines(N) [52]; Sri Lanka(N) [52].
Description[52,54].
Environmental[52]; Forage[52]; Toxins[52].

var. **siamensis** (Hoss.)Gagnep. [52,54]

I. spicata var. *siamensis* (Hoss.)De Kort & Thijsse [52,54].
Not climbing [52]; Herb or Shrub[52]; Perennial[52].
Indo-China: Thailand(N) [52]; Vietnam(N) [52].
Description[52].

I. hirsuta L. [29]

I. hirsuta var. *pumila* Bak. [52].
Not climbing [29]; Herb or shrub[29,52]; Perennial[29].
Indo-China: Cambodia(N) [29]; Laos(N) [29]; Thailand(N) [29]; Vietnam(N) [29]. Asia: Burma(N) [104]; China(N) [29]; India(N) [29]; Indonesia-ISO(N) [29]; Sri Lanka(N) [29]; Taiwan(N) [29].
Description[29,52]; Illustration[52].
Chemical Products[52]; Environmental[52]; Forage[52]; Medicine[52]; Toxins[52].
Widespread in the Old World; introduced in the New World. *I.astragalina* is sometimes synonymised with this, but is distinct in Africa [52].

I. kasinii Boonyamalik [53]
Not climbing [53]; Shrub[53]; Perennial[53].
Indo-China: Thailand(N) [53].
Description[53]; Illustration[53].
Endemic to Thailand [53].

I. kerrii De Kort & Thijsse [52]
Not climbing [52]; Shrub[52]; Perennial[52].
Indo-China: Thailand(N) [52].
Description[52].
Endemic to Thailand [52].

I. lacei Craib [52]
Not climbing [52]; Shrub[52]; Perennial[52].
Indo-China: Thailand(N) [52]. Asia: Burma(N) [52]; India(N) [104].
Description[52]; Illustration[52].

I. laxiflora Craib [52]
Not climbing [52]; Herb or Shrub[52]; Perennial[52].
Indo-China: Thailand(N) [52].
Description[52]; Illustration[52].
Endemic to Thailand [52].

I. linifolia (L.f.)Retz. [29]
Hedysarum linifolium [29].
Not climbing [29]; Herb[29]; Perennial[29].
Indo-China: Cambodia(N) [29]; Thailand(N) [29]; Vietnam(N) [29]. Asia: Burma(N) [29]; China(N) [29]; India(N) [29]; Indonesia-ISO(N) [29]; Sri Lanka(N) [29]; Taiwan(N) [29].
Description[29,52]; Distribution Map[52]; Illustration[52].
Food or Drink[52]; Medicine[52]; Toxins[52].
Widespread from NE Africa to Australia (Schrire, pers.comm.).

I. linnaei Ali [29]
I. dominii Eichler [52]; *I. enneaphylla* L. [29].
Not climbing [29]; Herb or shrub[29,52]; Perennial[29].
Asia: Burma(N) [29]; India(N) [29]; Indonesia-ISO(N) [29]; Sri Lanka(N) [29]. Indo-China: Laos(N) [29]; Thailand(N) [29]; Vietnam(N) [29].
Description[29]; Illustration[52].
Food or Drink[52]; Forage[52]; Medicine[52]; Toxins[52].
Widespread from Pakistan to Australia (Schrire, pers.comm.).
Causes 'Birdsville disease' in horses in Australia [52].

I. longicauda Thuan [29]
Not climbing [29]; Shrub[29]; Perennial[29].
Indo-China: Vietnam(N) [29].
Description[29,52]; Illustration[29].
Endemic to Vietnam [29].

I. nigrescens King & Prain [29]
Not climbing [29]; Shrub[29]; Perennial[29].
Indo-China: Laos(N) [29]; Thailand(N) [52]; Vietnam(N) [29]. Asia: Burma(N) [29]; China(N) [29]; India(N) [29]; Indonesia-ISO(N) [29]; Taiwan(N) [29].
Description[29]; Illustration[52].

I. nummulariifolia (L.)Alston [29]
I. echinata Willd. [29]. *Hedysarum nummulariifolium* L. [29].
Not climbing [29]; Herb or shrub[29,52]; Perennial[29].
Indo-China: Cambodia(N) [29]; Thailand(N) [29]; Vietnam(N) [29]. Asia: India(N) [29]; Sri Lanka(N) [29].
Description[29,52]; Illustration[29,52].

I. reticulata Franchet [52]
Not climbing [52]; Herb or Shrub[52]; Perennial[52].
Indo-China: Thailand(N) [52]. Asia: China(N) [52].
Description[52].

I. sootepensis Craib [29]

Not climbing [52]; Shrub[52]; Perennial[52].
Indo-China: Cambodia(N) [52]; Laos(N) [29]; Thailand(N) [52]; Vietnam(N) [52].
Description[29,52].

subsp. **sootepensis** [52]

Not climbing [52]; Shrub[52]; Perennial[52].
Indo-China: Cambodia(N) [52]; Laos(N) [29]; Thailand(N) [52]; Vietnam(N) [52].
Description[29,52].

subsp. **acutifolia** De Kort & Thjisse [29]

Not climbing [52]; Shrub[52]; Perennial[52].
Indo-China: Thailand(N) [52].
Description[52].

I. squalida Prain [29]

I. changensis Craib [52]; *I. polygaloides* Gagnep. [29].
Not climbing [29]; Herb or Shrub[29]; Perennial[29].
Indo-China: Cambodia(N) [29]; Laos(N) [29]; Thailand(N) [29]; Vietnam(N) [29]. Asia: Burma(N) [29]; China(N) [29].
Description[29,52]; Illustration[29,52].

I. stachyodes Lindl. [104]

I. dosua var. *tomentosa* Bak. [104].
Not climbing[104]; Herb or Shrub[104]; Perennial[104].
Indo-China: Cambodia(N) [104]; Thailand(N) [104]; Vietnam(N) [104]. Asia: Burma(N) [104]; China(N) [104]; India(N) [104]; Nepal(N) [104].
Description[104].
Nguyen van Thuan et al.(1987; ref.29) treat this as a synonym of *I. dosua.*

I. suffruticosa Miller [29,52]

Not climbing [52]; Shrub[52]; Perennial[52].
Indo-China: Cambodia(N) [29]; Laos(N) [29]; Thailand(N) [29]; Vietnam(N) [29]. Asia: China(N) [29]; India(N) [29]; Indonesia-ISO(N) [29]; Philippines(N) [29]; Taiwan(N) [29].
Description[29,52]; Illustration[52].
Chemical Products[52]; Environmental[52]; Medicine[52]; Toxins[52].
Often cultivated as a dye plant. Subsp. *guatemalensis* is a native of tropical America, cultivated in Java, but not naturalised [52].

subsp. **suffruticosa** [52]

I. anil L. [29].
Not climbing [29]; Shrub[29]; Perennial[29].
Indo-China: Cambodia(N) [29]; Laos(N) [29]; Thailand(N) [29]; Vietnam(N) [29]. Asia: China(N) [29]; India(N) [29]; Indonesia-ISO(N) [29]; Philippines(N) [29]; Taiwan(N) [29].
Description[29,52]; Illustration[52].
Chemical Products[52]; Environmental[52]; Medicine[52] Toxins[52].
Often cultivated as a dye plant.

I. tinctoria L. [29]

I. bergii Vatke [52]; *I. indica* Lam. [52]; *I. sumatrana* Gaertn. [52]; *I. tinctoria* Blanco [52]; *I. tinctoria* var. *torulosa* Bak.f. [52].
Not climbing [29]; Shrub[29]; Perennial[29].
Indo-China: Cambodia(N) [29]; Thailand(N) [29]; Vietnam(N) [29]. Asia: Burma(N) [29]; China(N) [29]; India(N) [29]; Indonesia-ISO(N) [29]; Japan(N) [29]; Malaysia-ISO(N) [29]; Pakistan(N) [29]; Sri Lanka(N) [29]; Taiwan(N) [29].
Description[29].
Chemical Products[29]; Environmental[29]; Medicine[29]; Toxins[29].
Pantropical distribution [29].

I. trifoliata L. [29]

Not climbing [52]; Herb or Shrub[52]; Perennial[52].
Indo-China: Laos(N) [29]; Thailand(N) [29]; Vietnam(N) [29]. Asia: Burma(N) [29]; China(N) [29]; India(N) [29]; Indonesia-ISO(N) [29]; Philippines(N) [29]; Taiwan(N) [29].
Description[29,52]; Distribution Map[52]; Illustration[52].
Food or Drink[52].
Only the typical subspecies is found in Indo-China; subsp. *unifoliolata* occurs in the Philippines [52].

subsp. **trifoliata** [52]

I. barberi Gamble [52]; *I. karuppiana* Pallithanan [52]; *I. trifoliata* var. *brachycarpa* Gagnep. [29].
Not climbing [29]; Herb[29]; Perennial[29].
Indo-China: Laos(N) [29]; Thailand(N) [29]; Vietnam(N) [29]. Asia: Burma(N) [29]; China(N) [29]; India(N) [29]; Indonesia-ISO(N) [29]; Philippines(N) [29]; Taiwan(N) [29].
Description[29,52]; Distribution Map[52]; Illustration[52].
Food or Drink[52].

I. trita L.f. [29]

Not climbing [52]; Shrub[52]; Perennial[52].
Indo-China: Laos(N) [52]; Thailand(N) [52]. Africa. Asia. Australasia.
Description[52].

subsp. **scabra** (Roth)De Kort & Thijsse [52]

I. laotica Gagnep. [52]; *I. rutschuruensis* De Wild. [52]; *I. scabra* Roth [52]; *I. subulata* var. *scabra* (Roth)Meikle [52].
Not climbing [52]; Shrub[52]; Perennial[52].
Indo-China: Laos(N) [52]; Thailand(N) [52]. Asia: India(N) [52]; Sri Lanka(N) [52]. Africa. Indian Ocean: Madagascar(N) [52]. South America.
Description[52].

I. wightii Wight & Arn. [29]

I. inamoena Thwaites [29]; *I. pallida* Craib [52].
Not climbing [29]; Shrub[29]; Perennial[29].
Indo-China: Cambodia(N) [29]; Laos(N) [29]; Thailand(N) [29]; Vietnam(N) [29]. Asia: Burma(N) [104]; India(N) [29]; Sri Lanka(N) [29].
Description[29]; Illustration[52].

I. zollingeriana Miq. [29]

I. benthamiana Hance [29]; *I. teysmannii* Miq. [29].
Not climbing [29]; Shrub or Tree[29]; Perennial[29].
Indo-China: Laos(N) [29]; Thailand(N) [52]; Vietnam(N) [29]. Asia: China(N) [29]; Indonesia-ISO(N) [29]; Taiwan(N) [29].
Description[29]; Distribution Map[52]; Illustration[52].
Environmental[29].

MILLETTIEAE

Generic delimitation in the *Millettieae* is controversial and in active development. Geesink (Scala Millettiearum — Leiden Botanical series 8 (1984)) made numerous suggestions for a new and more logical classification. Unfortunately he did not make the necessary nomenclatural combinations and his scheme is only slowly being adopted as genera are revised. The larger genera, particularly *Derris* and *Millettia* will undoubtedly be subdivided in the near future. This listing is therefore somewhat provisional and will certainly soon become out-of-date.

AFGEKIA Craib

Three species, all shrubs or small trees, occurring in SE Asia.

A. filipes (Dunn)Geesink [61]

Adinobotrys filipes Dunn [61]; *Padbruggea filipes* (Dunn)Craib [32].
Indo-China: Thailand(N)[61]. Asia: China(N)[61].
Climbing[61]; Shrub[61]; Perennial[61].
Description[62].

A. mahidolae Burtt & Chermsirivathana [59]
Indo-China: Thailand(N)[59].
Climbing[59]; Shrub[59]; Perennial[59].
Description[59],Illustration[59].

A. sericea Craib [60]
Indo-China: Thailand(N)[60].
Climbing[60]; Shrub[60]; Perennial[60].
Description[60],Illustration[61].

AGANOPE Miq.

Woody climbers with about seven species. The genus occurs from Africa to southeast Asia.

A. heptaphylla (L.)Polhill [57]
Derris diadelpha (Blanco)Merr. [57]; *Derris exserta* Craib [57]; *Derris heptaphylla* (L.)Merr. [57]; *Derris sinuata* Thwaites [57].
Climbing[4]; Shrub[4]; Perennial[4].
Indo-China: Thailand(N)[57]. Asia: Burma(N)[57]; China(N)[57]; India(N)[57]; Sri Lanka(N)[57].
Santisuk 688: 'climber' [4].

A. thyrsiflora (Benth.)Polhill [57]
A. floribunda Miq. [57]; *A. macrophylla* Miq. [57]; *A. subavenis* Miq. [57]; *Derris eualata* Bedd. [57]; *Derris latifolia* Prain [57]; *Derris platyptera* Bak. [57]; *Derris pyrrothyrsa* Miq. [57]; *Derris thyrsiflora* (Benth.)Benth. [57]; *Derris thyrsiflora* var. *eualata* (Bedd.)Thoth. [57]; *Derris thyrsiflora* var. *wallichii* (Prain)Thoth. [57]; *Derris wallichii* Prain [57]; *Millettia thyrsiflora* Benth. [57].
Climbing[4]; Shrub[4]; Perennial[4].
Indo-China: Thailand(N)[4]; Vietnam(N)[4]. Asia: India(N)[57]; Indonesia-ISO(N)[57]; Malaysia-ISO(N)[57], Borneo[57].
Description[47].
Kerr 15825: 'climber, Thailand'. *Tsang* 27193: 'Vietnam' [4].

ANTHEROPORUM Gagnep.

Two species, both trees, endemic to Indo-China.

A. harmandii Gagnep. [58]
Not climbing[58]; Tree[58]; Perennial[58].
Indo-China: Vietnam(N)[58].
Description[58].

A. pierrei Gagnep. [58]
Not climbing[58]; Tree[58]; Perennial[58].
Indo-China: Thailand(N)[58]; Vietnam(N)[58].
Description[58].

DERRIS Lour.

A genus mainly of woody climbers, with 60–70 species mainly in southeast Asia. Some yield insecticidal compounds (rotenones), which are also highly toxic to fish. Revision of the genus, in progress, may lead to the recognition of some of the segregates discussed by Geesink (Ref. 61).

D. alborubra Hemsley [47]
Climbing or not[47]; Shrub or tree[47]; Perennial[47].
Asia: China(N)[47]. Indo-China: Laos(N)[47]; Thailand(N)[32]; Vietnam(N)[47].
Description[47].

D. amoena Benth. [32]
Climbing[68]; Shrub[68]; Perennial[68].
Indo-China: Thailand(N)[32]; Vietnam(N)[4]. Asia: Burma(N)[32]; Singapore(N)[32].
Description[68].
Tsang 30265: 'Vietnam' [4].

var. **amoena** [68]
Climbing[68]; Shrub[68]; Perennial[68].
Indo-China: Thailand(N)[32]; Vietnam(N)[4]. Asia: Burma(N)[32].
Description[68].

var. **maingayana** Prain [68]
Climbing[68]; Shrub[68]; Perennial[68].
Indo-China: Thailand(N)[32]. Asia: Singapore(N)[32].
Description[68].
This was treated as a full species, *D. maingayana* (Prain)Baker in the Flora of British India.

D. balansae Gagnep. [47]
Not climbing[47]; Shrub[47]; Perennial[47].
Indo-China: Vietnam(N)[47].
Description[47].

D. dalbergioides Bak. [47]
D. microphylla Hassk. [47].
Not climbing[47]; Tree[47]; Perennial[47].
Indo-China: Thailand(N)[32]; Vietnam(N)[47]. Asia: Burma(N)[32]; Indonesia-ISO(N)[47]; Malaysia-ISO(N)[32].
Description[47].
Wood[47].

D. elegans Benth. [69]
Climbing[69]; Shrub[69]; Perennial[69].
Indo-China: Thailand(N)[32]. Asia: Burma(N)[69]; India(N)[69]; Indonesia-ISO(N)[69]; Malaysia-ISO(N)[69]; Philippines(N)[69].
Description[69], Illustration[69].

var. **elegans** [32,69]
Climbing[69]; Shrub[69]; Perennial[69].
Indo-China: Thailand(N)[32]. Asia: Burma(N)[69]; India(N)[69]; Indonesia-ISO(N)[69]; Malaysia-ISO(N)[69]; Philippines(N)[69].
Description[69],Illustration[69].

var. **vestita** Prain [32]
Indo-China: Thailand(N)[32].
Climbing[69]; Shrub[69]; Perennial[69].
Description[68].

D. elliptica (Roxb.)Benth. [32,69]
Climbing[69]; Shrub[69]; Perennial[69].
Indo-China: Cambodia(N)[69]; Laos(N)[47]; Thailand(N)[69]; Vietnam(N)[47]. Asia: Burma(N)[69]; India(N)[69]; Indonesia-ISO(N)[69]; Malaysia-ISO(N)[69]; Philippines(N)[69].
Description[47,69].
Toxins[69].

var. **elliptica** [32,69]
Climbing[69]. Shrub[69]; Perennial[69].
Indo-China: Cambodia(N)[69]; Thailand(N)[69]; Vietnam(N)[47]. Asia: Burma(N)[69]; India(N)[69]; Indonesia-ISO(N)[69]; Malaysia-ISO(N)[69]; Philippines(N)[69].
Description[69].
Derris root[69]; Tuba Root[69].
Roots are a source of rotenone, used in the preparation of a valuable insecticide [69].

var. **malacensis** Prain [32,47,69]
Climbing[69]; Shrub[69]; Perennial[69].
Indo-China: Vietnam(N)[47].
Description[47].

var. **tonkinensis** Gagnep. [32,47,69]
Climbing[69]; Shrub[69]; Perennial[69].
Indo-China: Laos(N)[47]; Thailand(N)[47]; Vietnam(N)[47].
Description[47].

D. ferruginea (Roxb.)Benth. [32,69]
Climbing[69]; Shrub[69]; Perennial[69].
Indo-China: Laos(N)[47]; Thailand(N)[32]. Asia: Burma(N)[69]; India(N)[69].
Description[69].
Toxins[69].
Indian Tuba Root[69].
Sanjappa (Ref.108) treats this as a synonym of *D. elegans.*

D. laotica Gagnep. [47]
Climbing[47]; Shrub[47]; Perennial[47].
Indo-China: Cambodia(N)[32]; Laos(N)[47]; Thailand(N)[47].
Description[47],Illustration[47].

var. **laotica** [47]
Climbing[47]; Shrub[47]; Perennial[47].
Indo-China: Cambodia(N)[32]; Laos(N)[47]; Thailand(N)[47].
Description[47],Illustration[47].

var. **virens** Gagnep. [47]
Climbing[47]; Shrub[47]; Perennial[47].
Indo-China: Cambodia(N)[47]; Laos(N)[32]; Thailand(N)[32].
Description[47].

D. marginata (Roxb.)Benth. [32,69]
Climbing[69]; Shrub[69]; Perennial[69].
Indo-China: Thailand(N)[32]; Vietnam(N)[47]. Asia: Burma(N)[69]; India(N)[69].
Description[69].

D. monticola (Kurz)Prain [32,69]
Millettia monticola Kurz [69].
Climbing[69]; Shrub[69]; Perennial[69].
Indo-China: Thailand(N)[32]. Asia: Burma(N)[69];India(N)[69].
Description[69],Illustration[69].

D. polyphylla Koord. & Val. [47]
Not climbing[47]; Tree[47]; Perennial[47].
Indo-China: Laos(N)[47]; Vietnam(N)[47]. Asia: Indonesia-ISO(N)[47].
Description[47].

D. reticulata Craib [60]
Climbing[60]; Shrub[60]; Perennial[60].
Indo-China: Thailand(N)[60].
Description[60].

D. robusta (DC.)Benth. [32,69]
Not climbing[69]; Tree[69]; Perennial[69].
Indo-China: Thailand(N)[32]. Asia: Burma(N)[69]; China(N)[69]; India(N)[69].
Description[69].
Environmental[69]; Forage[69]; Wood[69].

D. scandens (Roxb.)Benth. [32,69]
D. timoriensis (DC.)Pittier [69].
Climbing[69]; Shrub[69]; Perennial[69].
Indo-China: Cambodia(N)[47]; Laos(N)[47]; Thailand(N)[69]; Vietnam(N)[47]. Asia: Burma(N)[69]; China(N)[69]; India(N)[69]; Indonesia-ISO(N)[69]; Malaysia-ISO(N)[69]; Sri Lanka(N)[69].
Description[69],Illustration[69].
Fibre[69]; Toxins[69].
Hog Creeper[69].

D. thorelii Craib [32]

Millettia thorelii Gagnep. [32,47].
Climbing[47]; Shrub[47]; Perennial[47].
Indo-China: Laos(N)[32]; Thailand(N)[32].
Description[47].

D. tonkinensis Gagnep. [47]

Indo-China: Vietnam(N)[47]. Asia: China(N)[47].
Climbing or not[47]; Shrub or tree[47]; Perennial[47].
Description[47], Illustration[[47].

var. **tonkinensis** [47]

Indo-China: Vietnam(N)[47]. Asia: China(N)[47].
Climbing or not[47]; Shrub or tree[47]; Perennial[47].
Description[47], Illustration[47].

var. **compacta** Gagnep. [47]

Indo-China: Vietnam(N)[47].
Climbing or not[47]; Shrub or tree[47]; Perennial[47].
Description[47].

D. trifoliata Lour. [32,69]

D. heterophylla (Willd.)K.Heyne [69]; *D. uliginosa* (Willd.)Benth. [69].
Climbing[69]; Shrub[69]; Perennial[69].
Indo-China: Thailand(N)[32]; Vietnam(N)[69]. Asia: Burma(N)[69]; China(N)[69]; India(N)[69]; Indonesia-ISO(N)[69]; Malaysia-ISO(N)[69]; Sri Lanka(N)[69].
Description[69].

D. truncata Craib [60]

Indo-China: Thailand(N)[60].
Not climbing[60]; Shrub[60]; Perennial[60].
Description[60].

ENDOSAMARA Geesink

A recently segregated genus containing one or two species, both lianas, from Southeast Asia.

E. racemosa (Roxb.)Gees. [61]

Millettia leiogyna Kurz [70]; *Millettia pallida* Dalz. [70]; *Millettia racemosa* Benth. [61]; *Pongamia racemosa* Grah. [70]; *Robinia racemosa* Roxb. [70]; *Tephrosia racemosa* Wight & Arn. [70].
Climbing[70]; Shrub[70]; Perennial[70].
Indo-China: Thailand(N)[70]. Asia: Burma(N)[70]; India(N)[70]; Philippines(N)[70].
Description[61,70].

FORDIA Hemsley

A genus of about 10 species, mostly understorey treelets, often cauliflorous, found in southeast Asia and Malesia.

F. cauliflora Hemsley [64]

Millettia cauliflora Gagnep. [64].
Not climbing[64]; Tree[64]; Perennial[64].
Asia: China(N)[64].
Description[64].
Only known from the type, from China[64]. Gagnepain [47] records from Tonkin (Vietnam), presumably in error.

F. fruticosa Craib [63]

Not climbing[63]; Shrub[63]; Perennial[63].
Indo-China: Thailand(N)[63].
Description[63].
Probably a *Millettia* [63,64].

F. pauciflora Dunn [64]
Millettia cauliflora Prain [64].
Not climbing[64]; Tree[64]; Perennial[64].
Indo-China: Thailand(N)[64]. Asia: Malaysia-ISO(N)[64].
Description[64].

MILLETTIA Wight & Arn.

A large pantropical genus of shrubs, trees and lianas, with perhaps as many as 100 species although there is a great need for careful revisionary studies in many areas of the world. A few are cultivated for their decorative blue-purple flowers. Some of the species will probably be transferred to the genus *Callerya* as a result of revisionary work in progress.

M. acutiflora Gagnep. [32]
Not climbing[47]; Tree[47]; Perennial[47].
Indo-China: Thailand(N)[32].
Description[47].

M. bassacensis Gagnep. [47]
Perennial[47].
Indo-China: Laos(N)[47].
Description[47].

M. boniana Gagnep. [47]
Climbing or not[47]; Shrub or tree[47]; Perennial[47].
Indo-China: Vietnam(N)[47].
Description[47].

M. brandisiana Kurz [70]
Not climbing[70]; Tree[70]; Perennial[70].
Indo-China: Thailand(N)[70]. Asia: Burma(N)[70].
Description[70].

M. buteoides Gagnep. [47]
Not climbing[47]; Tree[47]; Perennial[47].
Indo-China: Laos(N)[32]; Thailand(N)[32].
Description[47], Illustration[47].
Food or Drink[32].

var. **buteoides** [32,47]
Not climbing[47]; Tree[47]; Perennial[47].
Indo-China: Laos(N)[32]; Thailand(N)[32].
Description[47],Illustration[47].

var. **siamensis** Craib [32]
Not climbing[47]; Tree[47]; Perennial[47].
Indo-China: Thailand(N)[32].
Description[32].
Food or Drink[32].
Holotype of this variety at Kew determined as *M. pendula* by R.Geesink in 1977 [4].

M. caerulea Bak. [70]
Climbing[70]; Shrub[70]; Perennial[70].
Indo-China: Thailand(N)[70]. Asia: Burma(N)[70].
Description[70].

M. cochinchinensis Gagnep. [47]
Climbing[47]; Shrub[47]; Perennial[47].
Indo-China: Vietnam(N)[47].
Description[47].

M. decipiens Prain [70]

Not climbing[70]; Tree; Perennial[70].
Indo-China: Thailand(N)[4]. Asia: Malaysia-ISO(N)[70].
Description[70].
Niyomdham et al. 1057: 'Thailand' [4].

M. dielsiana Diels [47]

M. bockii Harms [47]; *M. cinerea* sensu auctt. [47]; *M. cinerea* var. *yunnansis* Pampan. [70]; *M. duclouxii* Pampan. [47].
Climbing[47]; Shrub[47]; Perennial[47].
Indo-China: Laos(N)[47]; Vietnam(N)[47]. Asia: China(N)[47].
Description[47].
Food or Drink[47].

M. diptera Gagnep. [47]

Not climbing[47]; Tree[47]; Perennial[47].
Indo-China: Vietnam(N)[47].
Description[47].

M. dorwardii Coll. & Hemsley [70]

Climbing[70]; Shrub[70]; Perennial[70].
Indo-China: Thailand(N)[32]. Asia: Burma(N)[70].
Description[70].

M. eberhardtii Gagnep. [47]

Not climbing[47]; Tree[47]; Perennial[47].
Indo-China: Vietnam(N)[47].
Description[47].

M. erythrocalyx Gagnep. [32]

Not climbing[47]; Tree[47]; Perennial[47].
Indo-China: Cambodia(N)[32]; Laos(N)[32]; Thailand(N)[32].
Description[47].
A syntype, *Pierre* 1034(K), determined as *M. ovalifolia* by Geesink in 1977. Formal synonymy not published, however. No notes available on possible synonymy of varieties [4].

var. **erythrocalyx** [47]

Not climbing[47]; Tree[47]; Perennial[47].
Indo-China: Cambodia(N)[32]; Laos(N)[32]; Thailand(N)[32].
Description[47].

var. **fusca** Gagnep. [47]

Indo-China: Laos(N)[47].
Not climbing[47]; Tree[47]; Perennial[47].
Description[47].

M. eurybotrya Drake [70]

Climbing[70]; Shrub[70]; Perennial[70].
Indo-China: Laos(N)[47]; Vietnam(N)[70].
Description[70].

M. extensa Bak. [32]

M. auriculata Brandis [32,70]; *M. auriculata* var. *extensa* Craib [32]; *M. macrophylla* Kurz [32]; *Otosema extensa* Benth. [32].
Climbing[70]; Shrub[70]; Perennial[70].
Indo-China: Cambodia(N)[32]; Thailand(N)[32]; Vietnam(N)[32]. Asia: Burma(N)[32]; India(N)[32].
Description[70].

M. fallax Craib [63]

Not climbing[63]; Tree[63]; Perennial[63].
Indo-China: Thailand(N)[63].
Description[63].

M. foliolosa Gagnep. [47]
Climbing or not[47]; Shrub or tree[47]; Perennial[47].
Indo-China: Laos(N)[47].
Description[47].

M. glaucescens Kurz [72].
Not climbing[72]; Tree[72]; Perennial[72].
Indo-China: Thailand(N)[32]. Asia: Burma(N)[72].
Description[32,72].

var. **glaucescens** [32,72]
Not climbing[72]; Tree[72]; Perennial[72].
Asia: Burma(N)[72].
Description[72].

var. **siamensis** Craib [32]
Not climbing[72]; Tree[72]; Perennial[72].
Indo-China: Thailand(N)[32].
Description[32,72].

M. harmandii Gagnep. [47]
Not climbing[47]; Tree[47]; Perennial[47].
Indo-China: Laos(N)[47].
Description[47].

M. hemsleyana Prain [70]
M. decipiens Prain,p.p. [70].
Not climbing[70]; Tree[70]; Perennial[70].
Indo-China: Thailand(N)[32]. Asia: Malaysia-ISO(N)[70].
Description[70].

M. ichthyochtona Drake [70]
Not climbing[70]; Tree[70]; Perennial[70].
Indo-China: Thailand(N)[32]; Vietnam(N)[70].
Description[70].
Toxins[47].
Cultivated[47].

M. kangensis Craib [63]
Not climbing[63]; Tree[63]; Perennial[63].
Indo-China: Thailand(N)[63].
Description[63].

M. kityana Craib [63]
Climbing[63]; Shrub[63]; Perennial[63].
Indo-China: Thailand(N)[63].
Description[63].

M. laotica Gagnep. [47]
Not climbing[47]; Tree[47]; Perennial[47].
Indo-China: Laos(N)[47].
Description[47].

M. latifolia Dunn [70]
Not climbing[70]; Tree[70]; Perennial[70].
Indo-China: Thailand(N)[47]; Vietnam(N)[47].
Description[70].
It is questionable if this species is a tree [70].

M. leucantha Kurz [32,70]

M. pendula Baker [32,70].
Not climbing[70]; Tree[70]; Perennial[70].
Indo-China: Laos(N)[47]; Thailand(N)[70]. Asia: Burma(N)[70].
Description[70].
Wood[47].
Bentham's name was not validly published. *M.leucantha* is the earliest validly published name [32].

M. lucida Gagnep. [47]

Perennial[47].
Indo-China: Laos(N)[47].
Description[47].

M. macrostachya Coll. & Hemsley [32]

Indo-China: Thailand(N)[32]. Asia: Burma(N)[32].
Not climbing[4]; Tree[4]; Perennial[4].
Description[32].
Hambdanand & Sangkhachand 268: 'Tree' [4].

var. **macrostachya** [32]

Not climbing[4]; Tree[4]; Perennial[4].
Indo-China: Thailand(N)[32]. Asia: Burma(N)[32].
Description[32].

var. **tecta** Craib [32]

Indo-China: Thailand(N)[32]. Asia: Burma(N)[32].
Not climbing[4]; Tree[4]; Perennial[4].
Description[32].
Hambdanand & Sangkhachand 268: 'Tree' [4].

M. nana Gagnep. [47]

Not climbing[47]; Shrub[47]; Perennial[47].
Indo-China: Cambodia(N)[47].
Description[47], Illustration[47].

M. nigrescens Gagnep. [47]

Not climbing[47]; Tree[47]; Perennial[47].
Indo-China: Laos(N)[47].
Description[47].

M. obovata Gagnep. [47]

Perennial[47].
Indo-China: Vietnam(N)[47]. Asia: China(N)[47].
Description[47].

M. ovalifolia Kurz [70]

Not climbing[70]; Tree[70]; Perennial[70].
Indo-China: Cambodia(N)[47]; Thailand(N)[70]. Asia: Burma(N)[70].
Description[70].

M. pachycarpa Benth. [70]

Climbing[70]; Shrub or tree; Perennial[70].
Indo-China: Thailand(N)[70]. Asia: Burma(N)[70]; China(N)[70]; India(N)[70].
Description[70].

M. pachyloba Drake [70]

Climbing[70]; Shrub[70]; Perennial[70].
Indo-China: Vietnam(N)[70]. Asia: China(N)[70].
Description[70].

M. penduliformis Gagnep. [47]

Climbing[47]; Shrub[47]; Perennial[47].
Indo-China: Laos(N)[47]; Vietnam(N)[47].
Description[47].

M. penicillata Gagnep. [47]
Perennial[47].
Indo-China: Vietnam(N)[47].
Description[47].

M. pierrei Gagnep. [47]
Climbing[47]; Shrub[47]; Perennial[47].
Indo-China: Cambodia(N)[47].
Description[47].

M. piscidia (Roxb.)Wight [70]
Not climbing[70]; Tree[70]; Perennial[70].
Indo-China: Thailand(N)[4]. Asia: India(N)[70].
Description[70].
Toxins[70].
R.Geesink et al. 7942: 'Thailand' [4].

M. principis Gagnep. [47]
Perennial[47].
Indo-China: Laos(N)[47]; Vietnam(N)[47].
Description[47].
Laos or Vietnam [47].

M. pubinervis Kurz [70].
Indo-China: Cambodia(N)[47]; Laos(N)[47]; Thailand(N)[70]; Vietnam(N)[47]. Asia: Burma(N) [70].
Not climbing[70]; Tree[70]; Perennial[70].
Description[70].
Wood[47].

M. pulchra Kurz [47]
M. yunnanensis Pampan. [47]; *Mundulea pulchra* Benth. [47]; *Tephrosia pulchra* Colebr. [47]; *Tephrosia tutcheri* Dunn [47].
Climbing or not[47]; Shrub or tree[47]; Perennial[47].
Indo-China: Laos(N)[47]; Vietnam(N)[4]. Asia: Burma(N)[47]; China(N)[47]; India(N)[47].
Description[47]. *Squires* 856: 'Vietnam' [4].

M. reticulata Benth. [47]
M. cognata Hance [47]; *M. purpurea* Yatabe [47].
Not climbing[47]; Shrub[47]; Perennial[47].
Indo-China: Vietnam(N)[47]. Asia: China(N)[47]; Hong Kong(N)[47].
Description[47].
Schot (pers.comm.) believes the record for Vietnam to be incorrect.

M. rigens (Craib)Niyomdham [108]
Pueraria rigens Craib [108]
Climbing[83]; Shrub[83]; Perennial[83].
Indo-China: Thailand(N) [83].
Description[83]; Distribution Map[83]; Illustration[83].
Known only from the type [83].

M. sericea (Vent.)Benth. [32]
Climbing[47]; Shrub[47]; Perennial[47].
Indo-China: Thailand(N)[32]; Vietnam(N)[47]. Asia: Indonesia-ISO(N)[32]; Malaysia-ISO(N) [32].
Description[47].

M. setigera Dunn [47]
Not climbing[47]; Tree; Perennial[47].
Indo-China: Vietnam(N)[47].
Description[47].

M. speciosa Benth. [70]
Climbing[70]; Shrub[70]; Perennial[70].
Indo-China: Vietnam(N)[47]. Asia: China(N)[70]; Hong Kong(N)[70].
Description[70].

M. spireana Gagnep. [47]
Climbing[47]; Shrub[47]; Perennial[47].
Indo-China: Laos(N)[47]; Thailand(N)[32].
Description[47].

M. unijuga Gagnep. [47]
Perennial[47].
Indo-China: Laos(N)[47].
Description[47].

M. venusta Craib [63]
Not climbing[63]; Shrub[63]; Perennial[63].
Indo-China: Thailand(N)[63].
Description[63].

M. verruculosa Gagnep. [47]
Perennial[47].
Indo-China: Laos(N)[47].
Description[47].

M. xylocarpa Miq. [71]
Not climbing[4]; Tree[4]; Perennial[4].
Indo-China: Thailand(N)[4].
Description[71].
Niyomdham 1247: 'Thailand; tree' [4].

PADBRUGGEA Miq.

About 10 species, mostly lianes, found in southeast Asia. It seems likely that revision of generic limits (in progress) will lead to its incorporation in the genus *Callerya*.

P. atropurpurea Craib [32]
Adinobotrys atropurpureus Dunn [32]; *Millettia atropurpurea* Wall. [32]; *Pongamia atropurpurea* Wall. [32]; *Whitfordiodendron atropurpureum* (Wall.)Merr.,p.p. [65].
Not climbing[66]; Tree[66]; Perennial[66].
Indo-China: Laos(N)[47]; Thailand(N)[66]; Vietnam(N). Asia: Burma(N)[66]; Indonesia-ISO(N)[66]; Malaysia-ISO(N)[66].
Description[66],Illustration[66].
Geesink has put both *Whitfordiodendron* & *Padbruggea* in synonymy under *Callerya* Endl. As no combinations exist at specific level under *Callerya* the earliest name, *Padbruggea* is used [61].
Schot (pers.comm.) records this taxon from Vietnam.

P. eriantha (Benth.)Craib [32]
Adinobotrys erianthus Dunn [32]; *Millettia eriantha* Benth. [32]; *Whitfordiodendron erianthum* (Benth.)Merr. [65].
Climbing[66]; Shrub[66]; Perennial[66].
Indo-China: Thailand(N)[32]. Asia: Malaysia-ISO(N)[32].
Description[44].

P. pubescens Craib [32]
Millettia atropurpurea Ridl.,p.p. [32]; *Whitfordiodendron pubescens* (Craib)Burkill [67].
Not climbing[63]; Tree[63]; Perennial[63].
Indo-China: Laos(N)[32]; Thailand(N)[63].
Description[63].
Probably not distinct from *P. atropurpurea* (Schot, pers.comm.).

PONGAMIA Vent.

A genus with only a single species, a tree widespread on the coasts of southeast Asia and in the western Pacific and Australia.

P. pinnata (L.)Pierre [32]
P. glabra Vent. [32].
Not climbing[4]; Tree[4]; Perennial[4].
Indo-China: Thailand(N)[32]; Vietnam(N)[4]. Asia: India(N)[32].
Kerr 12819: 'Tree'. *Pierre* s.n.: 'Vietnam' [4].

SARCODUM Lour.

A small genus - perhaps 4 species - from southeast Asia and Japan. All lianas.

S. scandens Lour. [61]
Clianthus scandens (Lour.)Merr. [56].
Climbing[61]; Shrub[61]; Perennial[61].
Indo-China: Vietnam(N)[61]. Asia: Indonesia-ISO(N)[56]; Philippines(N)[56].
Description[61].

TEPHROSIA Pers.

A large genus, perhaps containing as many as 400 species, with its greatest diversity in Africa. Herbs, or soft-wooded shrubs. Some species contain substances toxic to insects and fish, and at least one species has been cultivated and introduced as a fish poison widely in the Indian Ocean region.

T. coccinea Wall. [47]
Not climbing[47]; Perennial[47].
Indo-China: Laos(N)[47]; Vietnam(N)[47]. Asia: Burma(N)[47]; India(N)[47].
Description[47].

T. kerrii J.R.Drumm. [32]
Not climbing[47]; Shrub[47]; Perennial[47].
Indo-China: Thailand(N)[32].
Description[32].

T. luzoniensis Vogel [73]
T. brachystachya sensu auctt. [73]; *T. coarctata* Miq. [73]; *T. confertiflora* Benth. [47,73]; *T. dichotoma* sensu auctt. [73]; *T. villosa* sensu auct. [73].
Not climbing[73]; Herb or Shrub[73]; Perennial[73].
Asia: Indonesia-ISO(N)[73]; Philippines(N)[73].
Description[73], Illustration[73].
Environmental[73].

T. pumila (Lam.)Pers. [73]
Not climbing[73]; Herb or Shrub[73]; Perennial[73].
Indo-China: Thailand(N)[73]. Asia: India(N)[73]; Indonesia-ISO(N)[73]; Malaysia-ISO(N)[73]; Philippines(N)[73]. Africa.
Description[73], Illustration[73].
Environmental[73]; Forage[73].
Pantropical[73].

subsp. **pumila** [73]
T. dichotoma Desv. [73]; *T. procumbens* (Ham.)Benth. [73]; *T. purpurea* var. *pumila* (Lam.)Bak. [73]; *T. timoriensis* DC. [73].
Not climbing[73]; Herb or Shrub[73]; Perennial[73].
Indo-China: Thailand(N)[73]. Asia: India(N)[73]; Indonesia-ISO(N)[73]; Philippines(N)[73]; Indian Ocean: Madagascar(N)[73]. Africa.
Description[73],Illustration[73].
Environmental[73]; Forage[73].

subsp. **aldabrensis** (J.R.Drumm. & Hemsley)Bosman & De Haas [73]
T. aldabrensis J.R.Drumm. & Hemsley [73]; *T. pumila* var. *aldabrensis* (J.R.Drumm. & Hemsley) Brummitt [73]; *T. pumila* var. *ciliata* (Craib)Brummitt [73]; *T. purpurea* var. *angustata* Miq. [73]; *T. purpurea* var. *ciliata* Craib [73].
Not climbing[73]; Herb or Shrub[73]; Perennial[73].
Indo-China: Thailand(N)[32]. Asia: Indonesia-ISO(N)[73]. Africa.
Description[73].

T. purpurea (L.)Pers. [47,73]
T. colonila (Ham.)Benth. [73]; *T. diffusa* (Roxb.)Wight & Arn. [73]; *T. purpurea* var. *diffusa* (Roxb.) Aitch. [73].
Not climbing[47]; Herb[47]; Perennial[47].
Indo-China: Cambodia(I)[47]; Laos(I)[47]; Thailand(I)[47]; Vietnam(I)[47]. Asia: China(I)[47]; India(N)[73]; Indonesia-ISO(N)[73]; Malaysia-ISO(I)[73]; Philippines(I)[73]; Sri Lanka(I) [73].
Description[73], Illustration[73].
Chemical Products[73]; Food or Drink[73]; Forage[73]; Medicine[73]; Toxins[73].
Native of India or Java [73].

subsp. **purpurea** [73]
T. wallichii Fawc. & Rendle [73]; *Cracca wallichii* (Fawc. & Rendle)Rydberg [73].
Not climbing[73]; Herb or Shrub[73]; Perennial[73].
Indo-China: Cambodia(I)[47]; Laos(I)[47]; Thailand(I)[47]; Vietnam(I)[47]. Asia: Indonesia-ISO(N)[73]; Malaysia-ISO(I)[73]; Philippines(I)[73]; Sri Lanka(I)[73].
Description[73], Illustration[73].
Chemical Products[73]; Food or Drink[73]; Forage[73]; Medicine[73]; Toxins[73].
Cultivated [73].

T. siamensis J.R.Drumm. [47]
T. coccinea Gagnep., p.p. [32,47].
Not climbing[47]; Shrub[47]; Perennial[47].
Indo-China: Thailand(N)[47]; Vietnam(N)[32].
Description[47].

T. vestita Vogel [73]
T. papuana Stemm. [73]; *T. repentina* Craib [73]; *T. tinctoria* sensu auct. [73].
Not climbing; Herb or Shrub[73]; Perennial[73].
Indo-China: Laos(N)[73]; Thailand(N)[73]; Vietnam(N)[73]. Asia: China(N)[73]; Indonesia-ISO(N)[73]; Malaysia-ISO(N)[73]; Philippines(N)[73].
Description[73], Illustration[73].

T. villosa (L.)Pers. [73]
Not climbing[73]; Herb[73]; Perennial[73].
Indo-China: Vietnam(N)[47]. Asia: India(N)[73]; Indonesia-ISO(N)[73]; Sri Lanka(N)[73]. Africa.
Description[73], Illustration[73].
Environmental[73]; Medicine[73].

PHASEOLEAE

AMPHICARPAEA Nuttall

A small herbaceous genus with one species on African mountains and two or perhaps more species in east Asia and North America. Species delimitation and the ranks accorded to the taxa are controversial.

A. edgeworthii Benth. [86]
A. trisperma Miq. [86]; *Shuteria anomala* Pamp. [86].
Climbing[86]; Herb[86]; Perennial[86].
Indo-China: Vietnam(N) [86]. Asia: China(N) [86]; India(N) [86].
Description[86]; Illustration[86].
Ohashi regards this as a subspecies of *A. bracteata* (L.)Fernald, a North American species.

A. ferruginea Benth. [98]

Shuteria siamensis Craib
Climbing[99]; Herb[99]; Perennial[99].
Indo-China: Thailand(N) [32]. Asia: Bhutan(N) [98]; Burma(N) [98]; China(N) [98]; India(N) [98]; Nepal(N) [32,98].
Description[32], Distribution Map[98].
Craib was uncertain of the genus of his new taxon; '*Galactia*; *Pueraria*; *Dumasia* ?' [32].
Ohashi has determined the type of *S. siamensis* (K) as *A. ferruginea* [100].
Nguyen van Thuan synonymised *S. siamensis* with *Amphicarpaea edgeworthii*, probably wrongly [86].
He also synonymised *A. ferruginea* with *Shuteria hirsuta* [86]; This is wrong, according to notes on material at Kew [99].
There is also *Shuteria ferruginea* (= *Pueraria ferruginea*, = *S. hirsuta*, q.v.) [100].

var. **ferruginea** [98]

A. edgeworthii var. *rufescens* Franchet [98]; *A. rufescens* (Franchet)Y.T.Wei & S.Lee [98]; *Shuteria ferruginea* (Benth.)Baker [98].
Climbing[98]; Herb[98]; Perennial[98].
Asia: Bhutan(N) [98]; Burma(N) [98]; China(N) [98]; Nepal(N) [98].
Distribution Map[98]; Illustration[98].

var. **bracteosa** (Prain)Ohashi & Tateishi [98]

Pueraria strobilifera Kurz [98]; *Shuteria bracteosa* Clarke [98]; *Shuteria ferruginea* var. *bracteosa* Prain [98].
Climbing[98]; Herb[98]; Perennial[98].
Asia: Burma(N) [101]; India(N) [98]. Indo-China: Thailand(N) [98].
Description[98],Distribution Map[98]; Illustration[98,101].

APIOS Fabr.

Ten herbaceous species in eastern Asia and North America. Some have edible tubers.

A. carnea (Wall.)Baker [86]

Cyrtotropis carnea Wall. [86].
Climbing[86]; Herb[86]; Perennial[86]
Indo-China: Thailand(N) [86]; Vietnam(N) [86]. Asia: Burma(N) [32]; China(N) [86]; India(N) [86]; Nepal(N) [86].
Description[47,86]; Illustration[86].

ATYLOSIA Wight & Arn.

A genus that is now generally regarded as forming part of *Cajanus* (q.v.).

A. trichodon Dunn [94]

Climbing[94]; Herb[94].
Indo-China: Thailand(N) [87]. Asia: China(N) [87].
Description[94].
Belongs to *Dunbaria*, but no combination available [87,103]; Wei & Lee have reduced this to the synonymy of *Atylosia barbata* (= *Cajanus goensis*, q.v.).[87,103].

BUTEA Willd.

A genus of four species of trees, occurring in south and southeast Asia. The species have many uses and are extremely decorative when in flower.

Butea monosperma (Lam.)Taubert [86]

B. frondosa Willd. [86]; *Erythrina monosperma* Lam. [86].
Not climbing[86]; Tree[86]; Perennial[86]. Indo-China: Cambodia(N); Laos(N); Thailand(N) [32]; Vietnam(N) [86]. Asia: Burma(N) [102]; India(N) [86]; Indonesia-ISO(N) [86]; Nepal(N) [102]; Sri Lanka(N) [86].
Description[47,86,102], Distribution Map[102]; Illustration[86,102].
Chemical Products[47].

Butea superba Roxb. [86]

Not climbing[86]; Tree[86]; Perennial[86].
Indo-China: Cambodia(N) [86]; Laos(N) [86]; Thailand(N) [32]; Vietnam(N) [86]. Asia: Burma(N) [102]; India(N) [102].
Description[47,86,102], Distribution Map[102]; Illustration[102].
Chemical Products[47].
Very close to *B. monosperma*; for distinctions see Sanjappa (1989) [86].

CAJANUS DC.

A genus of about 30 species of herbs, soft-wooded shrubs or climbers, most of which were formerly placed in the genus *Atylosia*. One species is an important and widely planted pulse crop. See Van der Maesen (1985 - ref. 83).

C. cajan (L.)Millsp. [87]

C. bicolor DC. [83]; *C. indicus* Spreng. [83]; *Cytisus cajan* L. [86].
Not climbing[83]; Shrub[83]; Perennial[83].
Indo-China: Cambodia(I) [86]; Laos(I) [86]; Thailand(I) [83]; Vietnam(I) [86].
Description[47,86,87], Distribution Map[87]; Illustration[86,87].
Domestic[83]; Environmental[83]; Food or Drink[83]; Forage[83]; Medicine[83].
Pigeon Pea [87].
Originated in India and is now cultivated in all tropical areas [83]. Cultivated as food plant for the lac insect (shellac) [47].

C. crassus (King)Maesen [87]

Climbing[87]; Shrub; Perennial[87].
Indo-China: Laos(N) [86]; Thailand(N) [87]; Vietnam(N) [87]. Asia: Burma(N) [87]; India(N) [87]; Indonesia-ISO(N) [87]; Malaysia-ISO(N) [87]; Nepal(N) [87]; Philippines(N) [87].
Description[47,86,87]; Distribution Map[87]; Illustration[86,87].
Food or Drink[86].
Two varieties occur. Var. *burmanicus* (Collett & Hemsley) Maesen & var. *crassus*. Only var. *crassus* occurs in S.E. Asia [87].

var. **crassus** [87]

Atylosia crassa King [87]; *Atylosia mollis* Benth., p.p. [87]; *Atylosia volubilis* (Blanco)Gamble [87] *Cantharospermum volubile* (Blanco)Merrill [87].
Climbing[87]; Shrub[87]; Perennial[87].
Indo-China: Laos(U) [86]; Thailand(U) [87]; Vietnam(U) [87]. Asia: Indonesia-ISO(U) [87]; Malaysia-ISO(U) [87]; Philippines(U) [87].
Description[86,87]; Distribution Map[87]; Illustration[86,87].
Food or Drink[86].

var. **burmanicus** (Collett & Hemsley)Maesen [87]

Atylosia burmanica Collett & Hemsley [87].
Asia: Burma(N) [87]; China(N) [87].

C. elongatus (Benth.)Maesen [87]

Atylosia elongata Benth. [87]; *Cantharospermum elongatum* (Benth.) Raizada [87].
Climbing[87]; Herb[87]; Perennial[87].
Indo-China: Vietnam(U) [87]. Asia: Burma(U) [87]; India(U) [87]; Nepal(U) [87].
Description[86,87]; Distribution Map[87]; Illustration[86,87].

C. goensis Dalz. [87]

Atylosia barbata (Benth.)Baker [87]; *Atylosia goensis* (Dalz.)Dalz. [87]; *Atylosia siamensis* Craib [87]; *Cantharospermum barbatum* (Benth.)Koord. [87]; *Dunbaria barbata* Benth. [86]; *Dunbaria stipulata* Thuan [87]; *Dunbaria thorelii* Gagnep., p.p. [87]; *Endomallus pellitus* Gagnep. [87]; *Endomallus spirei* Gagnep. [87].
Climbing[87]; Herb or shrub[87]; Perennial[87].
Indo-China: Laos(U) [86]; Thailand(U) [86]; Vietnam(U) [86]. Asia: Bangladesh(N) [104]; Burma(U) [87]; China(U) [87]; India(U) [87]; Indonesia-ISO(U) [87]; Malaysia-ISO(U) [87].
Description[86,87]; Distribution Map[87]; Illustration[47,86,87].
Food or Drink[86]; Medicine[87]

C. heynei (Wight & Arn.)Maesen [87]

Atylosia kulnensis (Dalz.)Dalz. [87]; *Dunbaria heynei* Wight & Arn. [87].
Climbing[87]; Shrub[87]; Perennial[87].
Indo-China: Vietnam(U) [105]. Asia: India(U) [87]; Sri Lanka(U) [87].
Description[87]; Distribution Map[87]; Illustration[87].
Van der Maesen did not see the Vietnamese material [87].

C. scarabaeoides (L.)Thouars [87]

Climbing or not[87]; Herb[87]; Perennial[87].
Indo-China: Cambodia(U) [86]; Laos(U) [86]; Thailand(U) [87]; Vietnam(U) [86]. Asia: Burma(U) [86]; China(U) [86]; India(U) [86]; Indonesia-ISO(U) [86]; Malaysia-ISO(U) [87]; Philippines(U) [86].
Description[47,86,87]; Distribution Map[87]; Illustration[86,87].
Environmental[87]; Forage[87]; Medicine[87]
Two varieties occur. Var. *pedunculatus* (Reynolds & Pedley)Maesen occurs in Australia. Var. *scarabaeoides* is the most widely distributed wild *Cajanus* [87].

var. **scarabaeoides** [87]

Atylosia scarabaeoides (L.)Benth. [87]; *Atylosia villosa* Bak. [86]; *Cantharospermum scarabaeoides* (L.)Baill. [87]; *Dolichos scarabaeoides* L. [87].
Climbing or not[87]; Herb[87]; Perennial[87].
Indo-China: Cambodia(U) [86]; Laos(U) [86]; Thailand(U) [87]; Vietnam(U) [86]. Asia: Burma(U) [86]; China(U) [86]; India(U) [86]; Indonesia-ISO(U) [86]; Malaysia-ISO(U) [87]; Philippines(U) [86].
Description[86,87]; Distribution Map[87]; Illustration[86,87].
Environmental[87]; Forage[87]; Medicine[87].

CALOPOGONIUM Desv.

Eight species of climbing herbs native to the tropics of the New World. One species is very widely planted throughout the tropics as forage and as a cover crop.

C. mucunoides Desv. [86]

Climbing[86]; Herb[86]; Perennial[86].
Indo-China: Laos(I) [86]; Vietnam(I) [86].
Description[86]; Illustration[86].
Environmental[86].
Native of Central & S.America; widely introduced in Old World Tropics [86].

CANAVALIA DC.

About 50 species, mostly large herbaceous climbers, throughout the tropics but most diverse in the New World. Some provide edible seeds.

C. africana Dunn [32]

C. virosa (Roxb.)Wight & Arn., p.p. [162]; *C. virosa* sensu Sauer [162,163].
Indo-China: Thailand(N) [32]. Asia: Burma(N) [32]; India(N) [32].
Climbing[32]; Herb[32]; Annual or perennial[32].
Kerr 1532b,cited as *C. virosa* by Craib, is *C. gladiata* (q.v.) [100]; *Winit* 1558, also cited by Craib, appears to be this but may be *C. ensiformis* [100]. Further study is needed to determine which cultivated taxa occur.

C. cathartica Thouars [86]

C. ensiformis (L.)DC. var. *turgida* Baker [162]; *C. microcarpa* (DC.)Piper [86]; *C. turgida* Gray [86]; *C. virosa* (Roxb.)Wight & Arn., p.p. [162]; *Dolichos virosus* [162].
Climbing[86]; Herb[86]; Perennial[86].
Indo-China: Cambodia(N) [86]; Thailand(N) [100]; Vietnam(N) [86]. Asia: India(N) [104].
Description[47,86]; Distribution Map[163]; Illustration[47].

C. ensiformis (L.)DC. [86]

Dolichos ensiformis L. [86].
Climbing[86]; Herb[86]; Perennial[86].
Indo-China: Laos(I) [86]; Vietnam(I) [86]. Asia: India(I) [104].
Description[86].
Food or Drink[86].
Possibly of American origin [86].

C. gladiata (Jacq.)DC. [86]

C. ensiformis sensu Merrill [86,106]; *C. foureiri* G.Don [86]; *C. gladiolata* Sauer [86]; *Dolichos gladiatus* Jacq. [86].
Climbing[86]; Herb[86]; Perennial[86].
Indo-China: Cambodia(N) [86]; Laos(N) [86]; Thailand(N) [100]; Vietnam(N) [86]. Asia: India(N) [104].
Description[86]; Distribution Map[163]; Illustration[86].
Food or Drink[86].
Probably native in E. Asia; range extended by cultivation [86].

C. lineata (Thunb.)DC. [86]

Dolichos lineatus Thunb. [86].
Climbing[86]; Herb[86]; Perennial[86].
Indo-China: Cambodia(N) [86]; Vietnam(N) [86]. Asia: China(N) [86]; Malaysia-ISO(N) [86].
Description[86]; Distribution Map[163]; Illustration[86].

C. rosea (Sw.)DC. [86]

C. maritima (Aubl.) Thouars [86]; *C. obtusifolia* (Lam.)DC. [86]; *Dolichos roseus* Sw. [86].
Climbing or not[86]; Herb[86]; Perennial[86].
Indo-China: Cambodia(N) [86; Thailand(N) [100]; Vietnam(N) [86]. Asia: China(N) [86]; India(N) [86]; Indonesia-ISO(N) [86].
Description[86]; Distribution Map[163]; Illustration[47].
A pantropical sea-shore species.

CLITORIA L.

About 70 species, mostly herbaceous climbers, mostly native to tropical America. The flowers are held inverted. Some are ornamental and cultivated.

C. cordiformis Fantz [107]

Climbing[107]; Herb[107]; Perennial[107].
Indo-China: Thailand(N) [107].
Description[107]; Illustration[107].

C. hanceana Hemsley [86]

Not climbing[86]; Shrub[86]; Perennial[86].
Indo-China: Cambodia(N) [86]; Laos(N) [86]; Thailand(N) [86]; Vietnam(N) [86]. Asia: China(N) [86].
Description[47,86].
Medicine[86].
3 other varieties recognized by Fantz (notes on sheets at K) — probably unpublished [100].*

var. **hanceana** [86]

Not climbing[86]; Shrub[86]; Perennial[86].
Indo-China: Cambodia(N) [86]; Thailand(N) [86]; Vietnam(N) [86]. Asia: China(N) [86].
Description[86].

* see p.133

var. **laureola** Gagnep. [86]
Not climbing[86]; Shrub[86]; Perennial[86].
Indo-China: Cambodia(N) [86]; Laos(N) [86]; Thailand(N) [86]; Vietnam(N) [86].
Description[86].
Medicine[86].

C. javanica Miq.
Climbing[100]; Herb[100]; Perennial[100].
Indo-China: Thailand(N) [100].
Kerr 13388, *Kerr* 19770 (both K) [100].

C. laurifolia Poir. [86]
C. cajanifolia (Presl)Benth. [86].
Climbing[86]; Herb[86]; Perennial[86].
Indo-China: Thailand(I) [86]; Vietnam(I) [86]. Asia: Burma(I) [86].
Description[47,86]; Illustration[86].
Native of Tropical America [86].

C. linearis Gagnep. [86]
Herb[86]; Perennial[86].
Indo-China: Laos(N) [86].
Description[47,86]; Illustration[47].
Known only from Laos [86].

C. macrophylla Benth. [86]
Climbing[86]; Herb[86]; Perennial[86].
Indo-China: Cambodia(N) [86]; Thailand(N) [86]; Vietnam(N) [86]. Asia: Burma(N) [86]; India(N) [86].
Description[47,86].
Close to *C.hanceana* Hemsley (q.v.) [86]; Sanjappa does not list this for India. Fantz (notes at K) has distinguished two varieties which appear to be unpublished [100].*

C. mariana L. [86]
Climbing[86]; Herb[86]; Perennial[86].
Indo-China: Laos(U) [86]; Thailand(U) [86]; Vietnam(U) [86]. Asia: Burma(U) [86]; India(U) [86].
Description[47,86]; Illustration[86].
Described from America. Native to Asia ?[86].
Fantz (notes at K) has distinguished a var. *orientalis*. Publication not traced [100].

C. ternatea L. [86]
Climbing[86]; Herb[86]; Perennial[86].
Indo-China: Cambodia(N) [86]; Laos(N) [86]; Thailand(N) [86]; Vietnam(N) [86]. Asia: Burma(N) [86]; China(N) [86]; India(N) [104].
Description[47,86]; Illustration[86].
Food or Drink[86]; Medicine[86].

CRUDDASIA Prain

A genus of one or perhaps two species, confined to southeast Asia. Two species transferred to this genus by Niyomdham have been retained in *Ophrestia* (q.v.) pending more detailed studies.

Cruddasia craibii Niyomdham [108]
Pueraria hirsuta sensu Craib [32,108].
Climbing[108]; Herb[108]; Annual or Perennial[108].
Indo-China: Thailand(N) [108].
Description[108]; Illustration[108].

Cruddasia insignis Prain [32]
Climbing[109]; Herb[109]; Perennial[109].
Indo-China: Thailand(N) [32]. Asia: Burma(N) [32]; India(N) [109].
Description[109,110]; Illustration[110].

* see p. 133.

DIOCLEA Kunth

Large climbers, mainly native to the New World tropics. A few species occur widely in the Old World, with seeds dispersed by sea drift. Specific limits and nomenclature of the Old World taxa are in doubt.

D. javanica Benth. [111]

Climbing[111]; Shrub[111]; Perennial[111].
Indo-China: Thailand(N) [100].
Description[111]; Illustration[111].
Verdcourt gives 'Indo-China' as part of distribution; this name is retained provisionally pending revision of the genus [111].

DIPHYLLARIUM Gagnep.

A poorly known monospecific genus from southeast Asia.

D. mekongense Gagnep. [86]

Indo-China: Laos(N) [86]; Vietnam(N) [86]. Asia: China(N) [86].
Not climbing[86]; Shrub[86]; Perennial[86].
Description[47,86]; Illustration[47,86].
Poorly known; relationships uncertain [100].

DOLICHOS L.

About 60 species in the Old World tropics. Most are climbing herbs, and most occur in the more seasonal vegetation types.

D. fragrans Kerr [112]

Climbing[112]; Herb[112]; Perennial[112].
Indo-China: Thailand(N) [112].
Description[112].

D. junghuhnianus Benth. [113]

D. henryi Harms [113].
Climbing[113]; Herb[113]; Perennial[113].
Indo-China: Thailand(N) [113]. Asia: China(N) [113]; Indonesia-ISO(N) [113].
Description[113].
Niyomdham has placed this as a synonym of *Vigna grahamiana* (Wight & Arn.)Verdc. [108]. Verdcourt (pers.comm.) does not agree.

D. tenuicaulis (Baker)Craib [86]

Climbing[86]; Herb[86]; Perennial[86].
Indo-China: Laos(N) [86]; Thailand(N) [32]. Asia: Burma(N) [104]; China(N) [104]; India(N) [104]; Nepal(N) [104].
Description[86]; Illustration[86].

subsp. **tenuicaulis** [86]

D. appendiculatus Hand.-Mazz. [86]; *D. falcatus* sensu auctt. [86]; *Phaseolus tenuicaulis* Baker [86].
Climbing[86]; Herb[86]; Annual or Perennial[86].
Indo-China: Thailand(N) [32].

subsp. **lygodioides** (Gagnep.)Ohashi & Tateishi [86]

D. lygodioides Gagnep. [86].
Climbing[86]; Herb[86]; Perennial[86].
Indo-China: Laos(N) [86].
Description[47,86]; Illustration[47,86].

D. thorelii Gagnep. [86]

D. balansae Gagnep. [86].
Climbing[86]; Herb[86]; Perennial[86].
Indo-China: Laos(N) [86]; Vietnam(N) [86].
Description[86].

D. trilobus L. [86]

D. debilis A.Rich. [86]; *D. falcatus* Willd.
Climbing[86]; Herb[86]; Perennial[86].
Indo-China: Laos(N) [86]; Vietnam(N) [86]. Asia: India(N) [104].
Description[86]; Illustration[86].
Taxonomy and relationships discussed at length by Verdcourt (1970) [113].

D. sp. Verdc. [113]

D. subcarnosus sensu Prain [113,114].
Climbing[113]; Herb[113]; Perennial[113].
Indo-China: Thailand(N) [113]. Asia: Bangladesh(N) [113]; Burma(N) [113].
Description[113].
Craib(1928) recorded *D.subcarnosus*; material at K redetermined by Verdcourt [100].
Niyomdham (1992) has placed this as a synonym of *Vigna grahamiana* (Wight & Arn.)Verdc. [108].

DUMASIA DC.

Eight species of climbing herbs occurring in the Old World tropics.

D. villosa DC. [86]

Climbing[86]; Herb[86]; Perennial[86].
Indo-China: Laos(N) [86]; Thailand(N) [86]; Vietnam(N) [86]. Asia: China(N) [86]; India(N) [86]; Indonesia-ISO(N) [86]; Malaysia-ISO(N) [32]; Nepal(N) [104]; Philippines(N) [32].
Description[47,86]; Illustration[86].

var. **villosa** [104]

D. glaucescens Miq. [86]; *D. pubescens* DC. [86].
Climbing[86]; Herb[86]; Perennial[86].
Indo-China: Laos(N) [86]; Thailand(N) [86]; Vietnam(N) [86].

var. **leiocarpa** (Benth.)Baker [104]

D. leiocarpa Benth. [104].
Climbing[86]; Herb[86]; Perennial[86].
Indo-China: Thailand(N) [32]. Asia: Burma(N) [32]; China(N) [104]; India(N) [32]; Nepal(N) [104].
Flora of Bhutan also treats this as a variety [100].

DUNBARIA Wight & Arn.

About 15 species of mainly climbing herbs found in southeast Asia, Malesia and Australasia. Revision in progress by Van der Maesen (WAG).

D. circinalis (Benth.)Baker [32]

Atylosia circinalis Benth. [87].
Climbing[87]; Herb[87]; Perennial[87].
Asia: Burma(N) [32]; India(N) [104]; Indonesia-ISO(N) [104]. Indo-China: Thailand(N) [32]; Vietnam(N) [95].
Craib was doubtful about this name but van der Maesen accepts it [87].

D. ferruginea Wight & Arn. [95].

Indo-China: Vietnam(N) [95].
A true *Dunbaria*, according to Van der Maesen [87].

D. flavescens Nguyen Van Thuan [86].
Climbing[86]; Herb[86]; Perennial[86].
Indo-China: Vietnam(N) [86].
Description[86,115]; Illustration[86,115].
Van der Maesen has annotated sheets at Kew as *D.crinita*(Dunn)van der Maesen.

D. fusca (Wall.)Kurz [86]
Climbing[86]; Herb[86]; Perennial[86].
Indo-China: Laos(N) [86]; Thailand(N) [86]; Vietnam(N) [86]. Asia: Burma(N) [86]; China(N) [165]; India(N) [86].
Description[47,86]; Illustration[86].
Not mentioned for India by Sanjappa (1992) [104].

var. **fusca** [86]
Atylosia crinita Dunn [87]; *Phaseolus fuscus* Wall. [86].
Climbing[86]; Herb[86]; Perennial[86].
Indo-China: Laos(N) [86]; Vietnam(N) [86].
Description[86].

var. **longicarpa** Nguyen Van Thuan [86]
Climbing[86]; Herb[86]; Perennial[86].
Indo-China: Laos(N) [86]; Vietnam(N) [86].
Description[86]; Illustration[86].

D. glabra Nguyen Van Thuan [86]
Climbing[86]; Herb[86]; Perennial[86].
Indo-China: Vietnam(N) [86].
Description[86]; Illustration[86].

D. glandulosa (Dalz. & Gibs.)Prain [32]
Atylosia glandulosa (Dalz. & Gibs.)Dalz. [87]; *Atylosia rostrata* Baker [32]; *C. glandulosus* Dalz. & Gibs. [32].
Climbing[32]; Herb[32]; Perennial[32].
Indo-China: Thailand(N) [32]. Asia: Burma(N) [32]; India(N) [32].
Description[47].

D. henryi Y.C.Wu [116]
Climbing[116]; Herb[116]; Perennial[116].
Indo-China: Vietnam(N) [116]. Asia: China(N) [116].
Description[116].

D. lecomtei Gagnep. [86]
Climbing[86]; Herb[86]; Perennial[86].
Indo-China: Vietnam(N) [86].
Description[47,86]; Illustration[47,86].

D. longiracemosa Craib [86]
D. longeracemosa Craib [86].
Climbing[86]; Herb[86]; Perennial[86].
Indo-China: Cambodia(N) [86]; Laos(N) [86]; Thailand(N) [86]; Vietnam(N) [86].
Description[47,86].
Van der Maesen (notes at K) has synonymized this with *D. bella* Prain [100].

D. nivea Miq. [86]
D. harmandii Gagnep. [86]; *D. scortechinii* Prain [86].
Climbing[86]; Herb[86]; Perennial[86].
Indo-China: Laos(N) [86]; Thailand(N) [32]; Vietnam(N) [86]. Asia: China(N) [86]; Indonesia-ISO(N) [86].
Description[47,86]; Illustration[47,86].
Food or Drink[86]; Toxins[86].
Type of *D.harmandii* at K annotated as *D.incana* (Zoll.& Mor.)van der Maesen [100]. *D. incana* may be the correct name for this taxon [100].

D. podocarpa Kurz [86]
Climbing[86]; Herb[86]; Perennial[86].
Indo-China: Cambodia(N) [86]; Laos(N) [86]; Thailand(N) [32]; Vietnam(N) [86]. Asia: Burma(N) [86]; China(N) [86]; India(N) [86].
Description[47,86]; Illustration[86].

D. punctata (Wight & Arn.)Benth. [100]
D. rhynchosioides Miq. [100].
Climbing[100]; Herb[100]; Perennial[100].
Indo-China: Thailand(N) [100]. Asia: Indonesia-ISO(N) [100].
Based on annotations by Van der Maesen at Kew, mainly on *Kerr* 1483b [100].

D. rotundifolia (Lour.)Merr. [86]
D. conspersa Benth. [86]; *Atylosia punctata* Dalz. [104]; *Indigofera rotundifolia* Lour. [86].
Climbing[86]; Herb[86]; Perennial[86].
Indo-China: Cambodia(N) [86]; Laos(N) [86]; Thailand(N) [32]; Vietnam(N) [86]. Asia: Bangladesh(N) [104]; Burma(N) [104]; China(N) [86]; India(N) [86]; Indonesia-ISO(N) [86]; Malaysia-ISO(N) [104]; Philippines(N) [104].
Description[47,86]; Illustration[86].

D. thorelii Gagnep. [86]
Indo-China: Laos(N) [86].
Climbing[86]; Herb[86]; Perennial[86].
Description[47,86].
Three out of four type sheets are *Cajanus goensis* (q.v.); judgement awaited on the fourth sheet which is a true *Dunbaria* [87].

D. truncata (Miq.) Maesen [87]
Climbing[87]; Herb[87]; Perennial[87].
Indo-China: Vietnam(N) [87].
Balansa 1229 cited as *D.subrhombea* by Nguyen Van Thuan 1979.[100]; named as *D. truncata* by van der Maesen at Kew.[100].

D. villosa (Thunb.)Makino [87]
Atylosia subrhombea Miq. [87]; *D. subrhombea* (Miq.)Hemsley [87].
Climbing[86]; Herb[86]; Perennial[86].
Indo-China: Cambodia(N) [86]; Laos(N) [86]; Thailand(N) [32]; Vietnam(N) [86]. Asia: China(N) [86]; Indonesia-ISO(N) [86].
Description[47,86]; Illustration[86].

DYSOLOBIUM (Benth.)Prain

Four species, all climbing herbs, from southeast Asia and Malesia. The limits of the genus are disputed as it occupies a rather intermediate position between *Dolichos* and *Vigna*.

D. apioides (Gagnep.)Maréchal [86]
Dolichos apioides Gagnep. [86].
Climbing[86]; Herb[86]; Perennial[86].
Indo-China: Cambodia(N) [86]; Laos(N) [86].
Description[47,86].
Van Welsen & den Hengst regard this as a synonym of *Dysolobium pilosum* (*Vigna pilosa*) [117].

D. dolichoides (Roxb.)Prain [86]
Climbing[86]; Herb[86]; Perennial[86].
Indo-China: Cambodia(N) [118]; Thailand(N) [86]; Vietnam(N) [86]. Asia: Bangladesh(N) [104]; India(N) [86]; Indonesia-ISO(N) [86].
Description[86,117]; Illustration[86].
Van Welzen & den Hengst do not recognize the varieties [117].

var. **dolichoides** [86]

Canavalia dolichoides (Roxb.)Kurz [86]; *Dolichos dasycarpus* Miq. [86]; *Phaseolus dolichoides* Roxb. [86]; *Vigna dolichoides* (Roxb.)Baker [86].
Climbing[86]; Herb[86]; Perennial[86].
Indo-China: Cambodia(N) [118]; Vietnam(N) [86]. Asia: India(N) [86]; Indonesia-ISO(N) [86].
Description[86]; Illustration[86].

var. **schomburgkii** (Gagnep)Maréchal [86]

Dolichos schomburgkii Gagnep. [86].
Climbing[86]; Herb[86]; Perennial[86].
Indo-China: Thailan(N) [86]; Vietnam(N) [118].

D. grande (Benth.)Prain [117]

Climbing[117]; Herb[117]; Perennial[117].
Indo-China: Thailand(N) [32]. Asia: Bangladesh(N) [104]; Burma(N) [32]; India(N) [32]; Nepal(N) [104].
Description[117]; Illustration[117].

D. pilosum (Willd.)Maréchal [86]

Dolichos pilosus Willd. [86]; *Dolichovigna formosana* Hayata [86]; *Dolichovigna pilosa* (Willd.) Niyomdham [108]; *Vigna pilosa* (Willd.)Baker [86].
Climbing[86]; Herb[86]; Perennial[86].
Indo-China: Cambodia(N) [86]; Laos(N) [86]; Thailand(N) [32]; Vietnam(N) [86]. Asia: Burma(N) [118]; India(N) [86]; Indonesia-ISO(N) [86]; Nepal(N) [118]; Philippines(N) [118].
Description[86,117]; Illustration[117].
Intermediate between *Vigna* and *Dysolobium* [100]; Van Welzen & den Hengst (1985) treat as *Vigna pilosa* [100]. Niyomdham (1992) treats as *Dolichovigna* [100].

ERIOSEMA (DC.)G.Don

A large genus, mainly of upright herbs and subshrubs, found mainly in Africa and the New World tropics.

E. chinense Vogel [86]

Not climbing[86]; Herb[86]; Perennial[86].
Indo-China: Cambodia(N) [86]; Laos(N) [86]; Thailand(N) [86]; Vietnam(N) [86]. Asia: Burma(N) [86]; China(N) [86]; India(N) [86]; Malaysia-ISO(N) [86].
Description[47,86]; Illustration[47,86].

ERYTHRINA L.

Trees, mostly spiny and with red flowers, found throughout the tropics but most diverse in the New World. Several species have been widely planted for ornament and as shade trees. About 110 species in all.

E. arborescens Roxb. [32]

Indo-China: Thailand(N) [32]. Asia: Bangladesh(N) [104]; Burma(N) [32]; China(N) [119]; India(N) [32]; Nepal(N) [119].
Not recorded for Thailand by Krukoff & Barneby (1974) [119].

E. fusca Lour. [86]

E. atrosanguinea Ridl. [86]; *E. glauca* Willd. [86]; *E. ovalifolia* Roxb. [86].
Not climbing[86]; Tree[86]; Perennial[86].
Indo-China: Cambodia(N) [86]; Laos(N) [86]; Thailand(N) [86]; Vietnam(N) [86]. Asia: Bangladesh(N) [104]; Burma(N) [86]; India(N) [86]; Indonesia-ISO(N) [86]; Philippines(N) [86]; Sri Lanka(N) [86].
Description[47,86]; Illustration[86].

E. stricta Roxb. [86]

Not climbing[86]; Tree[86]; Perennial[86].
Indo-China: Cambodia(N) [86]; Laos(N) [86]; Thailand(N) [32]; Vietnam(N) [86]. Asia: Bangladesh(N) [119]; Burma(N) [119]; China(N) [86]; India(N) [86]; Nepal(N) [119].
Description[47,86]; Illustration[86].

E. suberosa Roxb. [86]

E. alba Wight & Arn. [119]; *E. glabrescens* (Prain)R.N.Parker [119]; *E. stricta* Roxb. var. *suberosa* (Roxb.)Niyomdham [108]; *E. sublobata* Roxb. [119].
Not climbing[86]; Tree[86]; Perennial[86].
Indo-China: Cambodia(N) [86]; Thailand(N) [86]. Asia: Burma(N) [86]; China(N) [86]; India(N) [86].
Description[47,86].
Craib (1928) mentions a var. *horrida* Ridley, but is doubtful.
Niyomdham (1992) treats this as a variety of *E.stricta* (q.v.) [108].

E. subumbrans (Hassk.)Merr. [86]

E. holosericea Kurz,p.p. [119]; *E. lithosperma* Miq. [86]; *E. sumatrana* Miq. [119].
Not climbing[86]; Tree[86]; Perennial[86].
Indo-China: Laos(N) [86]; Thailand(N) [86]; Vietnam(N) [86]. Asia: Burma(N) [119]; India(N) [104]; Indonesia-ISO(N) [86]; Malaysia-ISO(N) [119]; Philippines(N) [86].
Description[47,86]; Illustration[86].

E. variegata L. [86]

E. indica Lam. [86]; *E. lithosperma* Hassk. [119]; *E. lobulata* Miq. [119]; *E. loureiri* G.Don [119]; *E. orientalis* (L.)Merr. [86]; *E. rostrata* Ridley [119]; *E. variegata* var. *orientalis* (L.)Merr. [86].
Not climbing[86]; Tree[86]; Perennial[86].
Indo-China: Cambodia(N) [86]; Laos(N) [86]; Thailand(N) [86]; Vietnam(N) [86]. Asia: China(N) [86]; India(N) [86]; Indonesia-ISO(N) [86]; Malaysia-ISO(N) [86]; Philippines(N) [86].
Description[47,86]; Illustration[47,86].
Domestic[86]; Environmental[86]; Food or Drink[86]; Medicine[86].

FLEMINGIA Wight & Arn.

About 30 species of shrubs and woody herbs of the Old World tropics, most diverse in southeast Asia. There is considerable divergence of taxonomic opinion and a thorough revision is much needed, covering the entire genus over its whole range. Some recent name changes based on regional studies have been mentioned here, but not adopted pending a full revision.

F. brevipes Craib [32]

Not climbing[120]; Shrub[120]; Perennial[120].
Indo-China: Thailand(N) [32].
Description[120].

F. chappar Benth. [86]

Moghania chappar (Benth.)O.Kuntze [86].
Not climbing[86]; Shrub[86]; Perennial[86].
Indo-China: Cambodia(N) [86]; Laos(N) [86]; Thailand(N) [86]. Asia: Burma(N) [86]; India(N) [86]; Nepal(N) [104].
Description[47,86]; Illustration[86].

F. ferruginea Benth. [86]

F. congesta var. *wightiana* (Wight & Arn.)Baker [86]; *F. ferruginea* var. *eglandulosa* Gagnep. [86].
Not climbing[86]; Shrub[86]; Perennial[86].
Indo-China: Laos(N) [86]; Thailand(N) [86]. Asia: Burma(N) [86]; India(N) [86]; Philippines(N) [86].
Description[47,86]; Illustration[86].
Sanjappa accepts *F.wightiana* Wight & Arn. as a good species endemic to India [104].
Nooteboom(1960) placed this as a synonym of *F.macrophylla (q.v.)* [121].

F. grahamiana Wight & Arn. [86]

F. pycnantha Benth. [104]; *F. sericans* Kurz [86]; *Moghania grahamiana* (Wight & Arn.)O.Kuntze [86].
Not climbing[86]; Shrub[86]; Perennial[86].
Indo-China: Laos(N) [86]; Thailand(N) [32]; Vietnam(N) [86]]. Asia: Burma(N) [86]; China(N) [86]; India(N) [86].
Description[47,86]; Illustration[86].
Sanjappa regards *F.sericans* as a synonym of *F.wallichii* Wight & Arn. [104].

F. involucrata Benth. [86]

F. capitata Zoll. [86]; *Moghania involucrata* (Benth.)O.Kuntze [86].
Not climbing[86]; Shrub[86]; Perennial[86].
Indo-China: Cambodia(N) [86]; Laos(N) [86]; Thailand(N) [86]; Vietnam(N) [86]. Asia: Bangladesh(N) [104]; Burma(N) [104]; India(N) [86]; Indonesia-ISO(N) [86]; Philippines(N) [86].
Description[47,86]; Illustration[86].

F. kerrii Craib [32]

Indo-China: Thailand(N) [32].
Description[47,122].

F. kradungensis Niyomdham [108]

Not climbing[108]; Shrub[108]; Perennial[108].
Indo-China: Thailand(N) [108].
Description[108]; Illustration[108].

F. lineata (L.)Ait.f.[86]

Not climbing[86]; Shrub[86]; Perennial[86].
Indo-China: Cambodia(N) [86]; Laos(N) [86]; Thailand(N) [86]; Vietnam(N) [86]. Asia: Burma(N) [86]; India(N) [86].
Description[47,86]; Illustration[86].
Food or Drink[86].

var. **lineata** [86]

Hedysarum lineatum L. [86]; *Moghania lineata* (L.)O.Kuntze [86].
Not climbing[86]; Shrub[86]; Perennial[86].
Indo-China: Cambodia(N) [86]; Laos(N) [86]; Vietnam(N) [86]. Asia: India(N) [86].
Description[86].

var. **glutinosa** Prain [86]

F. lineata var. *hirtella* Gagnep. [86].
Not climbing[86]; Shrub[86]; Perennial[86].
Indo-China: Laos(N) [86]; Thailand(N) [86]; Vietnam(N) [86]. Asia: Burma(N) [86].
Description[86]; Illustration[86].
Food or Drink[86].
Has been raised to specific rank by Wei & Lee (1985) [103].

F. macrophylla (Willd.)Prain [86]

F. congesta Ait.f.[86]; *F. congesta* var. *latifolia* Baker [86]; *F. congesta* var. *semialata* Baker [86]; *F. latifolia* Benth. [86]; *F. prostrata* Roxb. [104]; *Crotalaria macrophylla* Willd. [86]; *Moghania macrophylla* (Willd.)O.Kuntze [86].
Not climbing[86]; Shrub[86]; Perennial[86].
Indo-China: Cambodia(N) [86]; Laos(N) [86]; Thailand(N) [86]; Vietnam(N) [86]. Asia: Bangladesh(N) [104]; Burma(N) [86]; China(N) [86]; India(N) [86]; Nepal(N) [104].
Description[47,86]; Illustration[47,86].
Nooteboom (1960) takes a broad view of this species [121].

F. paniculata Benth. [86]

Moghania paniculata (Benth.)O.Kuntze [86].
Not climbing[86]; Shrub[86]; Perennial[86].
Indo-China: Laos(N) [86]; Thailand(N) [86]. Asia: Burma(N) [86]; India(N) [86]; Nepal(N) [104].
Description[86].

F. procumbens Roxb. [86]
F. vestita Baker [86].
Not climbing[86]; Shrub[86]; Perennial[86].
Indo-China: Laos(N) [86]; Vietnam(N) [86]. Asia: China(N) [86]; India(N) [86]; Philippines(N) [86].
Description[47,86].
Food or Drink[86].
Sometimes cultivated for the tubers, according to Sanjappa (1992) [104].
Placed as a synonym of *F.macrophylla* by Nooteboom (1960) [121].

F. sarmentosa Craib [32]
Climbing or not[60]; Shrub[60]; Perennial[60].
Indo-China: Thailand(N) [32].
Description[60].

F. sootepensis Craib [86]
F. macrophylla (Willd.)Prain var. *sootepensis* (Craib)Niyomdham [108].
Not climbing[86]; Shrub[86]; Perennial[86].
Indo-China: Laos(N) [86]; Thailand(N) [86]. Asia: China(N) [86].
Description[47,86]; Illustration[86].
Niyomdham (1992) treats this as a variety of *F. macrophylla* (q.v.) [108].

F. stricta Ait.f.[86]
Moghania stricta (Ait.)O.Kuntze [86].
Not climbing[86]; Shrub[86]; Perennial[86].
Indo-China: Cambodia(N) [86]; Laos(N) [86]; Thailand(N) [86]; Vietnam(N) [86]. Asia: Burma(N) [104]; China(N) [86]; India(N) [86]; Indonesia-ISO(N) [86]; Philippines(N) [86].
Description[47,86]; Illustration[86].
Domestic[86]; Food or Drink[86].
Filed with *F.macrophylla* (q.v.) at Kew [100].

F. strobilifera (L.)Ait.f.[86]
Not climbing[86]; Shrub[86]; Perennial[86].
Indo-China: Laos(N) [86]; Thailand(N) [86]; Vietnam(N) [86]. Asia: Bangladesh(N) [104]; Burma(N) [86]; China(N) [86]; India(N) [86]; Malaysia-ISO(N) [86]; Nepal(N) [104].
Description[47,86]; Illustration[86].
Medicine[86].

var. **strobilifera** [86]
F. bracteata Wight [86]; *F. strobilifera* var. *bracteata* Baker [86]; *Hedysarum strobiliferum* L. [86]; *Moghania strobilifera* (L.)O.Kuntze [86].
Not climbing[86]; Shrub[86]; Perennial[86].
Indo-China: Thailand(N) [86]; Vietnam(N) [86]. Asia: Burma(N) [86]; China(N) [86]; India(N) [86]; Malaysia-ISO(N) [86].
Description[86]; Illustration[86].
Medicine[86].

var. **fluminalis** (Prain)Nguyen Van Thuan [86]
F. fluminalis Prain [86].
Not climbing[86]; Shrub[86]; Perennial[86].
Indo-China: Laos(N) [86]; Vietnam(N) [86]. Asia: Burma(N) [86]; China(N) [86]; India(N) [86].
Description[47,86].

F. tiliacea Niyomdham [108]
Not climbing[108]; Herb or Shrub[108]; Perennial[108].
Indo-China: Thailand(N) [108]. Asia: India(N) [108].
Description[108]; Illustration[108].

F. wallichii Wight & Arn. [104]
Not climbing[104]; Herb[104]; Perennial[104].
Indo-China: Thailand(N) [100]. Asia: Burma(N) [104]; India(N) [104].

F. wangkae Craib [32]
Not climbing[120]; Shrub[120]; Perennial[120].
Indo-China: Thailand(N) [32].
Description[120].

GALACTIA P.Browne

About 50 species of twining herbs, sometimes woody, mainly in tropical America.

G. laotica Nguyen Van Thuan [86]
Climbing[86]; Herb[86]; Perennial[86].
Indo-China: Laos(N) [86]; Vietnam(N) [86].
Description[86]; Illustration[86].

G. latifolia (Baker)Nguyen Van Thuan [86]
G. tenuiflora var. *latifolia* Baker [86].
Climbing[86]; Herb[86]; Perennial[86].
Indo-China: Vietnam(N) [86]. Asia: India(N) [86].
Description[86]; Illustration[86].
Treated as a variety of *G. tenuiflora* by Sanjappa (1992) [104].

G. longipes Gagnep. [86]
G. mouretii Gagnep. [86].
Climbing[86]; Herb[86]; Perennial[86].
Indo-China: Laos(N) [86]; Vietnam(N) [86].
Description[47,86]; Illustration[47].

G. tenuiflora (Willd.)Wight & Arn. [86]
Glycine tenuiflora Willd [86].
Climbing[86]; Herb[86]; Perennial[86].
Indo-China: Thailand(N) [86]; Vietnam(N) [86]. Asia: China(N) [86]; India(N) [86]; Malaysia-ISO(N) [86].
Description[47,86]; Illustration[86].

G. vietnamensis Nguyen Van Thuan [86]
Indo-China: Cambodia(N) [86]; Vietnam(N) [86].
Climbing[86]; Herb[86]; Perennial[86].
Description[86,115]; Illustration[86,115].

GLYCINE Willd.

A genus of about nine species growing in east Asia and Australasia. One species is very widely grown for its edible seeds and the oil extracted from them.

G. max (L.)Merr. [86]
G. soja sensu auctt. [86]; *Soya max* (L.)Piper [32]; *Phaseolus max* L. [86].
Not climbing[86]; Herb[86]; Annual[86].
Indo-China: Laos(N) [86]; Thailand(N) [86]; Vietnam(N) [86].
Description[86]; Illustration[86].
Food or Drink[86].
Cultivated widely; unknown in the wild.; perhaps derived from the wild *Glycine soja* Sieb. & Zucc [86].

LABLAB Adans.

One species, a herbaceous twiner probably native only in Africa, where it is widespread. Varieties derived from this are cultivated throughout the Old World, though usually on a small scale.

L. purpureus (L.)Sweet [86]
Climbing or not[86]; Herb[86]; Annual or Perennial[86].
Indo-China: Cambodia(I) [86]; Laos(I) [86]; Thailand(I) [86]; Vietnam(I) [86].
Description[86]; Illustration[86].
Food or Drink[86].
Widely cultivated and sometimes naturalized.
Alternative classifications recognize cultivar groups.
The wild form (subsp. *uncinatus* Verdc.) is found only in Africa.

subsp. **purpureus** [86]
Dolichos lablab L.
Climbing or not[86]; Herb[86]; Annual or Perennial[86].
Indo-China: Cambodia(I) [86]; Laos(I) [86]; Thailand(I) [86]; Vietnam(I) [86].
Description[86]; Illustration[86].
Food or Drink[86].
Often cultivated and naturalized [86].

subsp. **bengalensis** (Jacq.)Verdc. [86]
Dolichos bengalensis Jacq. [86]; *Dolichos lablab* subsp. *bengalensis* (Jacq.)Rivals [86].
Not climbing[86]; Herb[86]; Annual or Perennial[86].
Indo-China: Laos(I) [86]; Vietnam(I) [86].
Description[86]; Illustration[86].
Food or Drink[86].
Widely cultivated [86].

MACROPTILIUM (Benth.)Urban

About 12 species of erect or twining herbs from the Neotropics. Some species have been introduced to the Old World tropics as fodder.

M. lathyroides (L.)Urban [86]
Not climbing[86]; Herb[86]; Annual[86].
Indo-China: Thailand(I) [32]; Vietnam(I) [86].
Description[86].

var. **semierectum** (L.)Urban [86]
Phaseolus lathyroides L.; *Phaseolus psoraleoides* Wight & Arn. [86]; *Phaseolus semierectus* L. [86].
Not climbing[86]; Herb[86]; Annual[86].
Indo-China: Thailand(I) [32]; Vietnam(I) [86].
Description[86]; Illustration[86].
Introduced from America [86].

MUCUNA Adans.

A genus of about 100 species of tropical climbers, often very large, some with decorative flowers in large pendent spikes. The pods of many species bear extremely irritant hairs; hairless forms are sometimes planted as fodder.

M. bracteata Kurz [123]
M. brevipes Craib [123]; *M. venulosa* (Piper)Merrill & Metcalf [124].
Climbing[123]; Herb or Shrub[123]; Perennial[123].
Indo-China: Laos(N) [123]; Thailand(N) [123]; Vietnam(N) [123]. Asia: Burma(N) [123]; China(N) [123].
Description[123]; Distribution Map[123]; Illustration[123].
Poilane 14551 from Cambodia is said to be this by Nguyen van Thuan (Ref. 86) but is not mentioned by Wilmot-Dear (Ref. 123).

M. gigantea (Willd.)DC. [123]
Climbing[123]; Herb[123]; Perennial[123].
Indo-China: Thailand(N) [123]; Vietnam(N) [123]. Asia: Burma(N) [123]; China(N) [123]; India(N) [123]; Japan(N) [123].
Description[86,123,124]; Illustration[125].

subsp. **gigantea** [123]
Dolichos giganteus Willd. [86].
Climbing[123]; Herb[123]; Perennial[123].
Indo-China: Thailand(N) [123]; Vietnam(N) [123]. Asia: Burma(N) [123]; India(N) [123]; Malaysia-ISO(N) [123].
Description[86,123,124].

M. gracilipes Craib [123]
Climbing[123]; Herb[123]; Perennial[123].
Indo-China: Thailand(N) [123].
Description[123]; Distribution Map[123]; Illustration[123].
Fruit unknown; assigned provisionally to subgenus *Stizolobium* [123].

M. hainanensis Hayata [126]
Climbing[123]; Herb or Shrub[123]; Perennial[123].
Indo-China: Thailand(N) [126]; Vietnam(N) [126]. Asia: Burma(N) [126]; China(N) [126]; India(N) [126]; Philippines(N) [126].
Description[123,126]; Illustration[126].

subsp. **hainanensis** Hayata [126]
M. nigricans sensu Tateishi & Ohashi [126]; *M. nigricans* var. *hainanensis* (Hayata)Wilmot-Dear [124,126]; *M. nigricans* var. *hongkongensis* Wilmot-Dear [124,126]; *M. suberosa* Gagnep. [126].
Climbing[126]; Herb or Shrub[126]; Perennial[126].
Indo-China: Thailand(N) [123]; Vietnam(N) [123]. Asia: China(N) [123].
Description[123,126].

subsp. **multilamellata** Wilmot-Dear [126]
M. imbricata Baker,p.p. [126]; *M. nigricans* sensu auctt. [126].
Climbing[126]; Herb or Shrub[126]; Perennial[126].
Asia: Burma(N) [126]; India(N) [126]; Philippines(N) [126].
Description[126]; Illustration[126].

M. interrupta Gagnep. [123]
M. nigricans var. *cordata* Craib [123].
Climbing[123]; Herb or Shrub[123]; Perennial[123].
Indo-China: Cambodia(N) [123]; Laos(N) [123]; Thailand(N) [123]; Vietnam(N) [123]. Asia: Burma(N) [123]; China(N) [123].
Description[123]; Illustration[123].
May also occur in NE India & Bhutan [123].

M. macrocarpa Wall. [123]
M. bodinieri Lev. [86]; *M. collettii* Lace [123].
Climbing[123]; Herb or Shrub[123]; Perennial[123].
Indo-China: Laos(N) [123]; Thailand(N) [123]; Vietnam(N) [123]. Asia: Burma(N) [123]; China(N) [123].
Description[86,123]; Illustration[123].

M. monosperma Wight [123]
Stizolobium monospermum (Wight)O.Kuntze [127].
Climbing[127]; Herb or Shrub[127]; Perennial[127].
Indo-China: Thailand(N) [123]. Asia: Bangladesh(N) [127]; Burma(N) [127]; India(N) [127]; Sri Lanka(N) [127].
Description[123,127]; Illustration[127].

M. oligoplax Wilmot-Dear [128]
Climbing[128]; Shrub[128]; Perennial[128].
Indo-China: Thailand(N) [128].
Description[128]; Illustration[128].

M. pruriens (L.)DC. [123]
Climbing[123]; Herb or Shrub[123]; Perennial[123].
Indo-China: Cambodia(N) [123]; Thailand(N) [123]; Vietnam(N) [123]. Asia: Burma(N) [123]; China(N) [123]; Malaysia-ISO(N) [123].
Description[123,127]; Illustration[123,127].

var. **pruriens** [123]
M. prurita Hook. [127]; *Dolichos pruriens* L. [86].
Climbing[123]; Herb or Shrub[123]; Perennial[123].
Indo-China: Cambodia(N) [123]; Thailand(N) [123]; Vietnam(N) [123]. Asia: Burma(N) [127]; Malaysia-ISO(N) [127].
Description[123]; Illustration[123].

var. **hirsuta** (Wight & Arn.)Wilmot-Dear [123]

M. hirsuta Wight & Arn. [123].
Climbing[123]; Herb or Shrub[[123]; Perennial[123].
Indo-China: Thailand(N) [123]; Vietnam(N) [123]. Asia: India(N) [123].
Description[123,127]; Illustration[127].

var. **utilis** (Wight)Burck [123]

M. capitata Wight & Arn. [127]; *M. cochinchinensis* (Lour.)A.Chev. [124]; *M. nivea* (Roxb.)Wight & Arn. [127]; *M. utilis* Wight [124].
Climbing[123]; Herb or Shrub[123]; Perennial[123].
Indo-China: Thailand(N) [123]; Vietnam(N) [123]. Asia: China(N) [124]; India(N) [127].
Description[123,124,127].
Environmental[124]; Forage[124].

M. revoluta Wilmot-Dear [123]

M. biplicata sensu auctt. [123]; *M. imbricata* var. *bispicata* Gagnep. [123]; *M. interrupta* Gagnep,p.p. [123]; *M. nigricans* sensu Nguyen Van Thuan,p.p. [86,123].
Climbing[123]; Herb or Shrub[123]; Perennial[123].
Indo-China: Cambodia(N) [123]; Laos(N) [123]; Thailand(N) [123]; Vietnam(N) [123]. Asia: China(N) [123].
Description[123]; Distribution Map[123]; Illustration[123].

M. stenoplax Wilmot-Dear [123]

Climbing[123]; Herb[123]; Perennial[123].
Indo-China: Thailand(N) [123]. Asia: Malaysia-ISO(N) [123].
Description[123]; Illustration[123].

M. thailandica Niyomdham & Wilmot-Dear [123]

Climbing[123]; Herb or Shrub[123]; Perennial[123].
Indo-China: Thailand(N) [123].
Description[123]; Illustration[123].

NOGRA Merr.

Three species of twining herbs; tropical Asia.

N. grahamii (Benth.)Merr. [86]

Grona grahamii Benth. [86].
Climbing[86]; Herb[86]; Perennial[86].
Indo-China: Laos(N) [86]; Thailand(N) [86] Asia: India(N) [86].
Description[47,86]; Illustration[47,86].

OPHRESTIA H.M.L.Forbes

A small genus of twining or erect herbs, originally described from Africa. Niyomdham has transferred the two southeast Asian species to *Cruddasia* (q.v.) but Verdcourt believes that the African and southeast Asian plants belong to the same genus and his placement is retained pending further detailed studies.

Ophrestia laotica (Gagnep.)Verdc. [86]

Glycine laotica Gagnep. [108]; *Cruddasia laotica* (Gagnep.)Niyomdham [108]; *Paraglycine laotica* (Gagnep.)F.J.Hermann [108].
Not climbing[86]; Herb or Shrub[86]; Perennial[86].
Indo-China: Laos(N) [86].
Description[47,86]; Illustration[47].
Known only from Laos [86]; Some Thai material at K may be this (e.g. *Kerr* 3246, 3276, 2671, 21509, 5475) [100].

Ophrestia pinnata (Merr.)Verdc. [86]

Glycine pinnata Merr. [108]; *Cruddasia pinnata* (Merr.)Niyomdham [108]; *Paraglycine pinnata* (Merr.)F.J.Hermann [108].
Climbing[86]; Herb or Shrub[86]; Perennial[86].
Indo-China: Vietnam(N) [86]. Asia: China(N) [86].
Description[86]; Illustration[86].

PACHYRHIZUS DC.

About six species growing naturally in the New World tropics. One species has edible tubers for which it is sometimes cultivated.

P. erosus (L.)Urban [86]

P. angulatus DC. [86]; *Dolichos erosus* L. [86]; *Dolichos bulbosus* L. [86].
Climbing[86]; Herb[86]; Perennial[86].
Indo-China: Cambodia(I) [129]; Laos(I) [129]; Thailand(I) [129]; Vietnam(I) [129]. Asia: Burma(I) [129]; China(I) [129]; India(I) [129]; Indonesia-ISO(I) [129] Malaysia-ISO(I) [129].
Description[86,129]; Illustration[86,129].
Food or Drink[86]; Medicine[86]; Toxins[86].
Originates from Central America; natural distribution mapped by Sorensen (1988) [129].

PARACALYX Ali

Six species of twining herbs found in the Old World tropics.

P. scariosus (Roxb.)Ali [104]

Cylista scariosa Roxb. [104].
Climbing[104]; Herb[104]; Perennial[104]
Indo-China: Thailand(N) [32]. Asia: Burma(N) [104]; India(N) [104]; Nepal(N) [104]; Pakistan(I) [104].

PHASEOLUS L.

A tropical American genus of about 50 species, of which several species are very widely cultivated in the tropical and temperate zones as pulse and green vegetable crops.

P. lunatus L. [86]

Dolichos tonkinensis Bui-Quang-Chieu [130]; *Phaseolus tunkinensis* Lour. [86].
Climbing[86]; Herb[86]; Perennial[86].
Indo-China: Thailand(I) [32]; Vietnam(I) [86].
Description[47,86]; Illustration[47,86].
Food or Drink[86].
Widely cultivated and naturalized [86].

P. vulgaris L. [86]

Climbing or not [86]; Herb[86]; Annual[86]
Indo-China: Laos(I) [86]; Thailand(I) [86]; Vietnam(I) [86].
Description[86]; Illustration[86]
Food or Drink[86]
Several varieties cultivated; sometimes naturalized. American origin [86].

PSOPHOCARPUS DC.

Nine species of sprawling or twining herbs, mostly in Africa, but with one species which is widely cultivated in southeast Asia and Malesia and whose wild origin is unknown.

P. scandens (Endl.)Verdc. [86]

P. longipedunculatus Hassk. [86]; *Diesingia scandens* Endl. [86].
Climbing[86]; Herb[86]; Perennial[86]
Indo-China: Vietnam(I) [86].
Description[86].
Cultivated in Saigon Botanic Garden. Native of Africa [86].

P. tetragonolobus (L.)DC. [86]

Dolichos tetragonolobus L. [86].
Climbing[86]; Herb[86]; Perennial[86]
Indo-China: Cambodia(U) [86]; Laos(U) [86]; Thailand(U) [32]; Vietnam(U) [86]. Asia: China(U) [86]; India(U) [86]; Indonesia-ISO(U) [86]; Philippines(U) [86].
Description[86]; Illustration[86].
Food or Drink[86].
Widely cultivated. Roots(tubers), leaves, flowers, green pods and seeds all eaten [86].

PUERARIA DC.

About 20 species, mainly in east and southeast Asia. Twining or creeping herbs, sometimes woody. One species widely planted as a forage and cover crop.

P. alopecuroides Craib [83]

Climbing[83]; Shrub[83]; Perennial[83].
Indo-China: Thailand(N) [83]. Asia: Burma(N) [83]; China(N) [83].
Description[83]; Distribution Map[83]; Illustration[83].

P. candollei Benth. [83]

Climbing[83]; Shrub[83]; Perennial[83].
Indo-China: Laos(N) [83]; Thailand(N) [83]. Asia: Bangladesh(N) [83]; Burma(N) [83]; India(N) [83].
Description[83]; Distribution Map[83]; Illustration[83].

P. imbricata Maesen [83]

Climbing[83]; Shrub; Perennial[83].
Indo-China: Laos(N) [83]; Thailand(N) [83].
Description[83]; Distribution Map[83]; Illustration[83].
Related to *P. lobata* [83].

P. maesenii Niyomdham [108]

Climbing[108]; Herb[108]; Perennial[108].
Indo-China: Thailand(N) [108].
Description[108]; Illustration[108].

P. mirifica Airy Shaw & Suvat. [83]

P. candollei Benth. var. *mirifica* (Airy Shaw & Suvat.) Niyomdham [108].
Climbing[83]; Shrub[83]; Perennial[83].
Indo-China: Thailand(N) [83].
Description[83]; Distribution Map[83].
Medicine[83]; Toxins[83]. Closely related to *P. candollei*; perhaps only a variety of this [83].

P. montana (Lour.)Merr. [92]

Climbing[83]; Shrub[83]; Perennial[83].
Indo-China: Laos(I) [83]; Thailand(I) [83]; Vietnam(I) [83]. Asia: China(U) [83]; India(U) [83]; Japan(I) [83].
Description[83]; Distribution Map[83]; Illustration[83].
Environmental[83]; Fibre[83]; Food or Drink[83]; Forage[83]; Medicine[83]; Weed[83].
Kudzu[83].
P. montana (Lour.)Merr., the most common and widespread species of the genus is very variable. Var. *lobata* is widely introduced in the tropics [83].

var. **montana** [92]

P. tonkinensis Gagnep. [83]; *Dolichos montana* Lour. [83].
Climbing[83]; Shrub[83]; Perennial[83].
Indo-China: Laos(I) [83]; Thailand(I) [83]; Vietnam(I) [83]. Asia: Burma(U) [83]; China(U) [83]; Hong Kong(U) [83]; Japan(U) [83]; Philippines(U) [83]; Taiwan(U) [83].
Description[83,86]; Distribution Map[83]; Illustration[83]

var. **chinensis** (Ohwi)Maesen [92]

P. lobata var. *chinensis* Ohwi [92]; *P. thomsonii* Benth. [83]; *P. triloba* sensu auct. [86].
Climbing[83]; Shrub[83]; Perennial[83].
Indo-China: Laos(I)[83]; Vietnam(I)[83]. Asia: Burma(U)[83]; China(U)[83]; Hong Kong(U) [83];India(U) Philippines(U)[83].
Description[83,86]; Distribution Map[83]; Illustration[83,86]

var. **lobata** (Ohwi)Maesen [92]

P. lobata (Willd.)Ohwi [92]; *P. lobata* var. *lobata* (Willd.)Ohwi [92]; *P. thunbergiana* Sieb & Zucc.)Benth. [83].
Climbing[83]; Shrub[83]; Perennial[83]
Indo-China: Thailand(U) [83]; Vietnam(U) [83]. Asia: China(U) [83]; Indonesia-ISO(N) [83]; Japan(U)[83]; Malaysia-ISO(U)[83] Philippines(U)[83]. Australasia: Australia(U) [83].
Description[83]; Distribution Map[83]; Illustration[83]
Environmental[83]; Fibre[83]; Food or Drink[83]; Forage[83]; Medicine[83]; Weed[83].
Kudzu[83].

P. peduncularis (Benth.)Benth. [86]

Derris bonatiana Pampan. [83]; *Neustanthus peduncularis* Benth. [86]; *P. peduncularis* var. *violacea* Franch. [86]; *P. yunnanensis* Franch. [83].
Climbing[83]; Shrub[83]; Perennial[83].
Indo-China; Vietnam(N)[86]. Asia: Burma(N)[83]; China(N)[83]; India(N)[83]; Nepal(N)[83] Pakistan(N)[83].
Description[83,86]; Distribution Map[83]; Illustration[83].

P. phaseoloides (Roxb.)Benth.[83]

Climbing[83]; Herb[83]; Perennial[83].
Indo-China; Cambodia(I)[86]; Laos(I)[86]; Thailand(I)[86]; Vietnam(I)[86]. Asia: Burma(I) [86]; China(I)[86]; India(I)[86]; Indonesia-ISO[86]; Philippines(I)[86].
Description[83,86]; Distribution Map[83]; Illustration[83,86]
Environmental[83]; Fibre[83] Food or Drink[83]; Forage[83]; Medicine[83].
Tropical Kudzu[83]
Very widely introduced in the tropics[83].

var. **phaseoloides** [83]

Dolichos phaseoloides Roxb. [86].
Climbing[83]; Herb[83]; Perennial[83].
Indo-China: Cambodia(N) [86]; Laos(N) [86]; Thailand(N) [86]; Vietnam(N) [86]. Asia: Bangladesh(N) [83]; Burma(N) [83]; China(N) [83]; India(I) [83]; Indonesia-ISO(N) [83]; Nepal(I) [83]; Philippines(N) [83]; Sri Lanka(I) [83]; Taiwan(N) [83].
Description[83,86]; Distribution Map[83]; Illustration[83,86]].
Environmental[83]; Fibre[83]]; Food or Drink[83]; Forage[83]; Medicine[83]].
Tropical Kudzu[83]

var. **javanica** (Benth.)Bak. [83]

Neustanthus javanicus Benth. [86]; *P. javanica (Benth.)Benth.* [86].
Climbing[83]; Herb[83]; Perennial[83].
Indo-China: Cambodia(N) [86]; Laos(I) [86]; Thailand(I) [83]; Vietnam(N) [86]. Asia: Brunei(N) [83]; Indonesia-ISO(N) [83]; Malaysia-ISO(N) [83]; Philippines(N) [83]; Sri Lanka(N) [83].
Description[83,86]; Distribution Map[83]; Illustration[83].
Environmental[83]; Fibre[83]; Food or Drink[83]; Forage[83]; Medicine[83].
Puero[83]; Tropical Kudzu[83].
Var. *javanica* is more vigorous & more widespread than var. *phaseoloides* [83].

var. **subspicata** (Benth.)Maesen [83]

Neustanthus subspicatus Benth. [83]; *P. subspicata* (Benth.)Benth. [83].
Climbing[83]; Herb[83]; Perennial[83].
Indo-China: Thailand(N) [83]. Asia: Bangladesh(N) [83]; Burma(N) [83]; India(N) [83].
Description[83]; Distribution Map[83]; Illustration[83].
Environmental[83]; Fibre[83]; Food or Drink[83]; Forage[83]; Medicine[83].
The common variety in N.E. India [83].

P. stricta Kurz [83]

P. brachycarpa Kurz [83]; *P. collettii* Prain [83]; *P. collettii* var. *siamica* (Craib)Gagnep. [83]; *P. hirsuta* Kurz [83]; *P. longicarpa* Thuan [83]; *P. siamica* Craib [83].
Climbing[83]; Shrub[83]; Perennial[83]
Indo-China: Laos(N) [86]; Thailand(N) [83]. Asia: Burma(N) [83]; China(N) [83].
Description[47,83,86]; Distribution Map[83]; Illustration[83,86].

P. wallichii DC. [83]

Climbing or not[83]; Shrub[83]; Perennial[83]
Indo-China: Thailand[83]. Asia: Bangladesh(U) [83]; Burma(U) [83]; China(U) [83]; India(U) [83]; Nepal(U) [83].
Description[83]; Distribution Map[83].
Environmental[83].

RHYNCHOSIA Lour.

A tropical genus of about 200 species. Most are small herbs, erect or climbing; some are large with somewhat woody stems.

R. acuminatissima Miq. [95]

Indo-China: Vietnam(N) [95]; Herb[95]

R. avensis Baker [32]

R. candicans Kurz,p.p. [32].
Climbing[32]; Herb[32]; Perennial[32].
Indo-China: Thailand(N) [32]. Asia: Burma(N) [32].
Description[114].

R. bracteata Baker [86]

Climbing[86]; Herb[86]; Perennial[86].
Indo-China: Laos(N) [86]; Thailand(N) [32]. Asia: Burma(N) [86]; India(N) [86].
Description[47,86]; Illustration[86].

R. calcicola Craib [32]

Not climbing[63]; Herb[63]; Perennial[63].
Indo-China: Thailand(N) [32].
Description[63].

R. distans Craib [32]

Climbing[63]; Shrub[63]; Perennial[63].
Indo-China: Thailand(N) [63].
Description[63].

R. harae Ohashi & Tateishi [131]

Climbing[132]; Herb[132]; Perennial[132].
Indo-China: Laos(N) [131]; Thailand(N) [131]. Asia: Bhutan(N) [131]; India(N) [131]; Nepal(N) [131].
Description[132]; Illustration[132]

subsp. **ovalifoliolata** Ohashi & Tateishi [131]

Climbing[131]; Herb[131]; Perennial[131].
Indo-China: Laos(N) [131]; Thailand(N) [131].
Description[131]; Illustration[131].

R. longipetiolata Hoss. [32]

Indo-China: Thailand(N) [32]. Description[47].

R. marcanii Craib [32]

Climbing[63]; Herb[63]; Annual[63].
Indo-China: Thailand(N) [32].
Description[63].
Life-span uncertain; probably annual [63].

R. minima (L.)DC. [86]

Dolichos minimus L. [86].
Climbing[86]; Herb[86]; Perennial[86].
Indo-China: Thailand(N) [32]; Vietnam(N) [86].
Description[86]; Illustration[86].
Pantropical[86]; Sanjappa (1992) records var. *nuda* (DC.)O.Kuntze from Thailand [104].

R. nummularia (L.)DC. [86]

Glycine nummularia L. [86].
Climbing[86]; Herb[86]; Perennial[86].
Indo-China: Vietnam(N) [86]. Asia: India(N) [86].
Description[86]; Illustration[86].
Sometimes cultivated (but not stated why) [86].

R. rufescens (Willd.)DC. [86]

Glycine rufescens Willd. [86].
Not climbing[86]; Shrub[86]; Perennial[86].
Indo-China: Cambodia(N) [86]; Thailand(N) [32]. Asia: Bangladesh(N) [104]; Burma(N) [32]; India(N) [86]; Indonesia-ISO(N) [86].
Description[47,86]; Illustration[86].

R. volubilis Lour. [86]

Indo-China: Vietnam(N) [86]. Asia: China(N) [86].
Climbing[86]; Herb[86]; Perennial[86].
Description[47,86]; Illustration[86].

SHUTERIA Wight & Arn.

Five species of twining herbs occurring in south and southeast Asia.

S. annamica Gagnep. [86]

Climbing[133]; Herb[133], Shrub; Perennial[133].
Indo-China: Vietnam(N) [86].
Description[47,86,133]; Illustration[47].
Known only from the type (*Bon* 627) [86].

S. hirsuta Baker [86]

Pueraria anabaptis Kurz [83]; *Pueraria ferruginea* Kurz [83]; *S. anabaptis* (Kurz)C.Y.Wu [83]; *S. ferruginea* (Kurz)Baker [83].
Climbing[133]; Herb[133]; Perennial[133].
Indo-China: Laos(N) [133]; Thailand(N) [133]; Vietnam(N) [133]. Asia: Burma(N) [133]; China(N) [133]; India(N) [133].
Description[47,86,133]; Illustration[86,133].
Food or Drink[86].

S. involucrata (Wall.)Wight & Arn. [133]

S. pampaniniana Hand.-Mazz. [133]; *S. vestita* var. *villosa* Pamp. [133]
Climbing[133]; Shrub or Herb[133]; Perennial[133].
Indo-China: Cambodia(N) [133]; Thailand(N) [133]; Vietnam(N) [133]. Asia: China(N) [133]; India(N) [133]; Indonesia-ISO(N) [133].
Description[86,133]; Illustration[86,133].
Recorded for Cambodia (Nguyen van Thuan 1972),but not mentioned in Nguyen van Thuan 1979 [133].
Ohashi (1988) distinguishes a var. *vestita* (Wight & Arn.)Ohashi [134].

var. **involucrata** [168]

S. vestita var. *involucrata* Baker [168]; *S.sinensis* Hemsley [168]; *Glycine involucrata* Wall. [168]; *S. involucrata* var. *sinensis* (Hemsl.)Niyomdham [108].
Climbing[133]; Shrub or Herb[133]; Perennial[133].
Indo-China: Cambodia(N) [133]; Thailand(N) [168]; Vietnam(N) [133]. Asia: Burma(N) [168]; China(N) [168]; India(N) [168]; Nepal(N) [168].
Description[86,133]; Illustration[86,133].

var. **glabrata** (Wight & Arn.)Ohashi [168]

S. densiflora Benth. [168]; *S.glabrata* Wight & Arn. [168]; *S. vestita* Wight & Arn. [168]; *S. vestita* var. *densiflora* (Benth.)Baker [168]; *S. vestita* var. *glabrata* (Wight & Arn.)Baker [168].
Climbing[133]; Herb or Shrub[133]; Perennial[133].
Indo-China: Thailand(N) [133]; Vietnam(N) [133]. Asia: Burma(N) [168]; China(N) [168]; India(N) [168]; Malaysia-ISO(N) [133]; Nepal(N) [168]; Sri Lanka(N) [168].
Description[86,133]; Illustration[86,133].

var. **villosa** (Pampan.)Ohashi [169]

S. vestita Wight & Arn. var. *villosa* Pampan. [168]; *S.pampaniniana* Hand.-Mazz. [168].
Climbing[133]; Herb or Shrub[133]; Perennial[133].
Asia: China(N) [168].

S. suffulta Benth. [133]

Climbing[133]; Herb[133], Shrub; Perennial[133].
Indo-China: Thailand(N) [133]. Asia: Burma(N) [133]; China(N) [133]; India(N) [133].
Description[47,133]; Illustration[133].

SINODOLICHOS Verdc.

A poorly known genus of two species of twining herbs, confined to southeast Asia. Niyomdham (Ref. 108) has transferred the two species to *Dolichos*, but he did not take into account all the characters used by Verdcourt in defining *Sinodolichos*.

S. lagopus (Dunn)Verdc. [113]

Dolichos lagopus Dunn [113].
Climbing[100]; Herb[100]; Perennial[100].
Indo-China: Thailand(N) [100]; Vietnam(N) [100]. Asia: China(N) [113].
Niyomdham (1992) places this in *Dolichos* [108].

S. oxyphyllus (Benth.)Verdc. [113]

Dolichos oxyphyllus (Benth.)Niyomdham [108]; *Galactia oxyphylla* Benth. [113]; *Teramnus flexilis* Benth. [32]; *Teramnus oxyphyllus* (Benth.)Kurz [113].
Climbing[100]; Herb[100]; Annual or Perennial[100].
Asia: Burma(N) [113].
Sanjappa (1992) treats *T.flexilis* as a good species, but does not mention the earlier name *T.oxyphylla* [104].

SPATHOLOBUS Hassk.

About 15 species of climbers, sometimes large and woody, confined to south and southeast Asia and Malesia.

S. acuminatus Benth. [135]

S. listeri Prain [135]; *S. pallidus* Craib [136]; *S. roseus* Prain [135]; *S. squamiger* Prain [135]; *Butea acuminata* (Benth.)Kurz [135]; *Butea listeri* (Prain)Blatter [135]; *Butea rosea* (Prain)Blatter [135]; *Butea squamigera* (Prain)Blatter [135].
Climbing[135]; Shrub[135]; Perennial[135].
Indo-China: Laos(N) [135]; Thailand(N) [135]; Vietnam(N) [135]. Asia: Burma(N) [135]; India(N) [135]; Malaysia-ISO(N) [135].
Description[135]; Distribution Map[135]; Illustration[135].

S. harmandii Gagnep. [135]

S. compar Craib [135]; *S. sinensis* Chun & T.Chen [135]; *Butea harmandii* (Gagnep.)Blatter [135].
Climbing[135]; Shrub[135]; Perennial[135].
Indo-China: Laos(N) [135]; Thailand(N) [135]; Vietnam(N) [135]. Asia: Burma(N) [135]; China(N) [135]; Malaysia-ISO(N) [135].
Description[47,86,135]; Distribution Map[135].
Illustration[47,86,135].

S. parviflorus (DC.)O.Kuntze [135]

S. roxburghii Benth. [135]; *S. roxburghii* var. *denudatus* Baker [135]; *S. roxburghii* var. *platycarpus* Baker [135]; *Butea parviflora* DC. [135].
Climbing or not[135]; Shrub or Tree[135]; Perennial[135].
Indo-China: Cambodia(N) [135]; Laos(N) [135]; Thailand(N) [135]; Vietnam(N) [135]. Asia: Burma(N) [135]; India(N) [135].
Description[47,86,135]; Distribution Map[135]; Illustration[86,135]
China, *fide* Craib (1928), but not mentioned from there by Ridder-Numan & Wirinidata (1985) [135].

S. pottingeri Prain [135]

S. balansae Gagnep. [135]; *S. biauritus* Wei [136]; *S. dimorphus* Craib [135]; *S. spirei* Gagnep. [135]; *S. varians* Dunn [136]; *Butea balansae* (Gagnep.)Blatter [135]; *Butea pottingeri* (Prain)Blatter [136]; *Butea spirei* (Gagnep.)Blatter [135]; *Butea varians* (Dunn)Blatter [136].
Climbing[135]; Shrub[135]; Perennial[135].
Indo-China: Laos(N) [135]; Vietnam(N) [135]. Asia: Burma(N) [135]; China(N) [135]; India(N) [135].
Description[47,86,135]; Distribution Map[135]; Illustration[135].

S. suberectus Dunn [135]

S. floribundus Craib [135]; *S. laoticus* Gagnep. [135]; *Butea laotica* (Gagnep.)Blatter [135]; *Butea suberecta* (Dunn)Blatter [135].
Climbing[135]; Shrub[135]; Perennial[135]
Indo-China: Laos(N) [135]; Thailand(N) [135]; Vietnam(N) [135]. Asia: China(N) [135]; India(N) [135].
Description[47,86,135]; Illustration[135].

TERAMNUS P.Browne

A small genus of herbs, some climbing, in the seasonal tropics of the Old and New Worlds. Some are very variable and have been much subdivided.

T. labialis (L.f.)Spreng. [86]

Glycine labialis L.f. [32].
Climbing[86]; Herb[86], Shrub; Perennial[86].
Indo-China: Cambodia(N) [86]; Laos(N) [86]; Thailand(N) [86]; Vietnam(N) [86]. Asia: India(N) [86]; Indonesia-ISO(N) [86].
Description[47,86]; Illustration[47,86].
Sanjappa(1992) records *T. mollis* (*T.labialis* var. *mollis* (Wight & Arn.)Baker from Burma [104].

TEYLERIA Backer

A small genus, close to *Pueraria*, occurring in southeast Asia and Malesia.

T. barbata (Craib)Maesen [83]

Pueraria barbata Craib [83].
Climbing[60]; Herb[60]; Perennial[60].
Indo-China: Thailand(N) [83].
Description[60]; Distribution Map[83]; Illustration[83].

T. koordersii (Backer)Backer [83]

Glycine hainanensis Merr. & Metcalf [83]; *Glycine koordersii* Backer [83].
Climbing[91]; Herb[91]; Perennial[91].
Indo-China: Vietnam(N) [83]. Asia: China(N) [83]; Indonesia-ISO(N) [83].
Description[91]; Distribution Map[83]; Illustration[83]

T. tetragona (Merr.)Maesen [83]

Pueraria tetragona Merr. [83].
Climbing[85]; Herb[85]; Perennial[85].
Indo-China: Thailand(N) [83]. Asia: Indonesia-ISO(N) [83]; Philippines(N) [83].
Description[85]; Distribution Map[83]; Illustration[83].

VIGNA Savi

A complex and difficult genus of about 150 species in the tropics of the Old and New Worlds. Most are herbaceous climbers, and several species are cultivated as fodder and pulse crops.

V. adenantha (G.F.Meyer)Maréchal et al. [86]

Phaseolus adenanthus G.F.Meyer [86].
Climbing[86]; Herb[86]; Perennial[86].
Indo-China: Cambodia(N) [86]; Laos(N) [86]; Thailand(N) [32]; Vietnam(N) [86].
Description[47,86]; Illustration[86].
Food or Drink[47].

V. angularis (Willd.)Ohwi & Ohashi [86]

Azukia angularis (Willd.)Ohwi [86]; *Dolichos angularis* Willd. [86]; *Phaseolus angularis* (Willd.) W.F.Wight [86].
Climbing or not[86]; Herb[86]; Annual[86].
Indo-China: Vietnam(N) [86]. Asia: China(N) [86].
Description[86]; Illustration[86].
Food or Drink[86].
Adzuki bean[86].
Poilane 2122 (K) named as var. *nipponensis* by Tateishi [100].

V. dalzelliana (O.Kuntze)Verdc. [118]

Phaseolus dalzellianus O.Kuntze [118]; *Phaseolus dalzellii* Cooke [118]; *Phaseolus pauciflorus* Dalzell [118].
Indo-China: Cambodia(N) [86]; Laos(N) [86]; Thailand(N) [86]; Vietnam(N) [86]. Asia: India(N) [86]; Philippines(N) [86].
Description[86]; Illustration[86].

var. **dalzelliana** [86]

Climbing or not[86]; Herb[86]; Annual[86].
Asia: India(N) [86]; Philippines(N) [86].
Description[86].
Very similar to *V.minima*(Roxb.)Ohwi & perhaps conspecific (Maréchal et al.) [118].

var. **elongata** Nguyen Van Thuan [86]]

Herb[86]; Annual[86].
Indo-China: Cambodia(N) [86]; Laos(N) [86]; Vietnam(N) [86].
Description[86]; Illustration[86].

V. glabrescens Maréchal et al. [118]

Phaseolus calcaratus var. *glaber* (Roxb.)Prain [118]; *Phaseolus glaber* Roxb. [118]; *Phaseolus mungo* var. *glaber* (Roxb.)Baker [118]; *V. radiata* var. *glabra* (Roxb.)Verdc. [118].
Climbing[86]; Herb[86]; Annual[86]
Indo-China: Vietnam(N) [86]. Asia: China(N) [86]; India(N) [86].
Description[86].
Cultivated [86].
A natural amphidiploid – the only case in *Vigna* [166].

V. grahamiana (Wight & Arn.)Verdc. [113]

Dolichos grahamianus (Wight & Arn.)Niyomdham [108]; *Dolichos subcarnosus* Wight & Arn.; *Phaseolus grahamianus* Wight & Arn. [113].
Climbing[86]; Herb[86]; Perennial[86].
Indo-China: Thailand(N) [108].
Niyomdham (1992) regards this as a *Dolichos* and adds *D.henryi* and *D.junghuhnianus* to the synonymy [108].
Maréchal [166] agrees with Niyomdham's synonymy but not with his generic placement.

V. hirtella Ridley [100]

Climbing[100]; Herb[100]; Perennial[100].
Indo-China: Laos(N) [100].
Specimen at K (*Poilane* 28392) determined by Tateishi.
Maréchal et al (1970) [100] were unable to opine on the validity of this taxon [100].

V. luteola (Jacq.)Benth. [86]

Dolichos luteolus Jacq. [86]; *Phaseolus luteolus* (Jacq.)Gagnep. [86].
Climbing[86]; Herb[86]; Perennial[86].
Indo-China: Laos(N) [86]; Thailand(N) [32]; Vietnam(N) [86].
Description[47,86]; Illustration[86].

V. marina (Burm.)Merr. [86]

Phaseolus marinus Burm. [32]; *Phaseolus obovatus* Gagnep. [86]; *V. lutea* (Sw.)A.Gray [86].
Climbing or not[86]; Herb[86]; Annual or Perennial[86].
Indo-China: Thailand(N) [32]; Vietnam(N) [86].
Description[47,86]; Illustration[86]
Pantropical seashore species [86].

V. minima (Roxb.)Ohwi & Ohashi [118]

Azukia minima (Roxb.)Ohwi [118]; *Phaseolus minimus* Roxb. [118].
Indo-China: Laos(U) [100]; Thailand(U) [100]; Vietnam(U) [100].

subsp. **minima** [118]

Climbing[118]; Herb[118]; Annual[118].
Indo-China: Laos(U) [100]; Thailand(U) [100].

V. mungo (L.)Hepper [86]

Azukia mungo (L.)Masamune [86]; *Phaseolus mungo* L. [86].
Climbing or not[86]; Herb[86]; Annual[86].
Indo-China: Vietnam(I) [86]. Asia: India(U) [86].
Description[86]; Illustration[86].
Food or Drink[86].
Black gram[86]; Urd[86].

V. prainiana Babu & Sharma [137]

Phaseolus calcaratus var. *gracilis Prain* [137].
Climbing[137]; Herb[137]; Annual[137].
Indo-China: Cambodia(U) [137]; Thailand(U) [137]; Vietnam(U) [137].
Description[137].
Tateishi (notes at K) treats as a subsp. of *V. minima* (q.v.) [100].

V. radiata (L.)R.Wilczek [86]
Not climbing[86]; Herb[86]; Annual[86].
Indo-China: Cambodia(U) [86]; Laos(U) [86]; Thailand(U) [32]; Vietnam(U) [86]. Asia: India(U) [86].
Description[86]; Illustration[86].

var. **radiata** [86]
Azukia radiata (L.)Ohwi [86]; *Phaseolus aureus* Roxb. [86]; *Phaseolus chanetii* (Levl.)Levl. [83]; *Phaseolus radiatus* L. [86]; *Pueraria chanetii* Levl. [83].
Not climbing[86]; Herb[86]; Annual[86].
Indo-China: Cambodia(U) [86]; Laos(U) [86]; Thailand(U) [32]; Vietnam(U) [86]. Asia: India(U) [86].
Description[86]; Illustration[86].
Food or Drink[86].
Golden gram[86]; Green gram[86]; Mung bean[86].

var. **grandiflora** (Prain)Niyomdham [108]
Phaseolus sublobatus Roxb. var. *grandiflorus* Prain [108].
Climbing[108]; Herb[108]; Annual[108].
Indo-China: Thailand(U) [108].
Description[108].
Raised to specific status by Tateishi (notes at Kew) [100].

var. **sublobata** (Roxb.)Verdc. [86]
Phaseolus sublobatus Roxb. [86]; *Phaseolus trinervius* Wight & Arn. [86]; *V. brachycarpa* Kurz [86].
Climbing[86]; Herb[86]; Annual[86].
Indo-China: Laos(N) [86]; Thailand(N) [100]; Vietnam(N) [86]. Asia: China(N) [86]; India(N) [86]; Indonesia-ISO(N) [86].
Description[86]; Illustration[86].
Wild and cultivated [86].

V. trilobata (L.)Verdc. [86]
Dolichos trilobatus L. [86]; *Phaseolus trilobatus* (L.)Schreb. [86]; *Phaseolus trilobus* sensu auctt. [86].
Climbing[86]; Herb[86]; Perennial[86]
Indo-China: Vietnam(N) [86]. Asia: Burma(N) [86]; China(N) [86]; India(N) [86].
Description[86]; Illustration[86].
Wild and cultivated [86].

V. umbellata (Thunb.)Ohwi & Ohashi [86]
Herb[86]; Annual[86].
Indo-China: Cambodia(N) [86]; Laos(N) [86]; Thailand(N) [100]; Vietnam(N) [86]. Asia: China(N) [86]; India(N) [86]; Malaysia-ISO(N) [86]; Philippines(N) [86].
Description[86].
Food or Drink[86].

var. **umbellata** [86]
Azukia umbellata (Thunb.)Ohwi [86]; *Dolichos umbellatus* Thunb. [86]; *Phaseolus calcaratus* Roxb. [86]; *Phaseolus pubescens* Bl. [86]; *V. calcarata* (Roxb.)Kurz [86].
Herb[86]; Annual[86].
Indo-China: Cambodia(N) [86]; Laos(N) [86]; Thailand(N) [32]; Vietnam(N) [86]. Asia: China(N) [86]; India(N) [86]; Philippines(N) [86].
Description[86].
Food or Drink[86].
Rice bean[86].

var. **gracilis** (Prain)Maréchal et al. [86]
Herb[86]; Annual[86].
Indo-China: Cambodia(N) [86]; Laos(N) [86]; Thailand(N) [32]; Vietnam(N) [86]. Asia: Indonesia-ISO(N) [32]; Malaysia-ISO(N) [86].
Description[86].
Phaseolus calcaratus Roxb. var. *gracilis* Prain is a synonym of this, *fide* Nguyen van Thuan [100]; but a synonym of *V.prainiana* Babu & Sharma according to them.

V. unguiculata (L.)Walpers [86]
Climbing or not[86]; Herb[86]; Annual or Perennial[86].
Indo-China: Cambodia(N) [86]; Laos(N) [86]; Thailand(N) [32]; Vietnam(N) [86].
Description[86]; Illustration[86].
Food or Drink[86].
Many forms and cultivar groups; several rival classification systems [86].
Not native to Asia; wild forms found only in Africa.

subsp. **unguiculata** [86]
V. sinensis (L.)Hassk. [86]; *V. sinensis* subsp. *sinensis* (L.)Hassk. [86]; *V. unguiculata* var. *unguiculata* (L.)Walpers [86]; *Dolichos unguiculatus* L.
Climbing or not[86]; Herb[86]; Annual or Perennial[86].
Indo-China: Cambodia(I) [86]; Laos(I) [86]; Thailand(I) [32]; Vietnam(I) [86].
Description[86]; Illustration[86].
Food or Drink[86].
Black eye[86]; Cowpea[86]; Haricot dolique[86]; Pois du Bresil[86].

subsp. **cylindrica** (L.)Verdc. [86]
Dolichos catjang Burm.f. [86]; *V. cylindrica* (L.)Skeels [86]; *V. sinensis* var. *catjang* (Burm.f.)Chiov. [86]; *V. sinensis* subsp. *cylindrica* (L.)Van Eseltine [86]; *V. unguiculata* subsp. *catjang* (Burm.f.)Chiov. [86]; *V. unguiculata* var. *catjang* (Burm.f.)Ohashi [86]; *V. unguiculata* var. *cylindrica* (L.)Ohashi [86].
Climbing or not[86]; Herb[86]; Annual or Perennial[86].
Indo-China: Cambodia(I) [86]; Laos(I) [86]; Vietnam(I) [86].
Description[86]; Illustration[86].
Food or Drink[86].
Catjang[86]; Dolique blanc[86]; Dolique noir[86]; Dolique pourpre[86]; Horse gram[86]; Katchang[86]

subsp. **sesquipedalis** (L.)Verdc. [86]
V. sinensis subsp. *sesquipedalis* (L.)Van Eseltine [86]; *Dolichos sesquipedalis* L. [86].
Climbing[86]; Herb[86]; Annual[86].
Indo-China: Vietnam(I) [86].
Description[86].
Food or Drink[86].
Asparagus bean[86]; Dolique asperge[86]; Haricot baguette[86]; Yard-long bean[86].

V. vexillata (L.)A.Rich. [86]
Climbing[86]; Herb[86]; Perennial[86].
Indo-China: Cambodia(N) [86]; Laos(N) [86]; Thailand(N) [32]; Vietnam(N) [86].
Description[86]; Illustration[86].

var. **vexillata** [86]
Phaseolus vexillatus L. [86].
Climbing[86]; Herb[86]; Perennial.
Indo-China: Laos(U) [86]; Vietnam(U) [86].
Description[86]; Illustration[86].
Environmental[86].

var. **angustifolia** (Schum. & Thonn.)Baker [86]
Dolichos stenophyllus Gagnep. [86]; *V. vexillata* var. *linearis* Craib [86].
Climbing[86]; Herb[86]; Perennial[86]
Indo-China: Cambodia(U) [86]; Laos(U) [86]; Thailand(U) [32]. Asia: India(U) [86].
Description[86]
Var. *angustifolia* as understood by Maréchal et al is mainly African [118]. They question whether var. *linearis* is really the same [118].

var. **macrosperma** Maréchal et al. [86]
Climbing[86]; Herb[86]; Annual or Perennial[86].
Indo-China: Laos(N) [86]; Vietnam(N) [86]. Asia: China(N) [86].
Description[86]; Illustration[86].

PSORALEEAE

CULLEN Medikus

About 35 species of herbs or small shrubs of the Old World tropics and sub-tropics. Formerly included in *Psoralea.*

C. corylifolius (L.)Medikus [74,75]

Psoralea corylifolia L. [29,74]; *Trifolium unifolium* Forssk. [139].
Not climbing[29]; Herb[29]; Annual[29].
Indo-China: Laos(I)[29]; Vietnam(I)[29]. Asia: Burma(N)[139]; China(I)[29]; India(N)[29]; Pakistan(I)[29]; Sri Lanka(I)[29].
Description[29]; Illustration[29,140].
Medicine[29].
Originally from India [29].

ROBINIEAE

SESBANIA Scop.

A genus of fast-growing herbs or soft wooded shrubs, sometimes attaining tree size. About 50 species in the warmer regions of the world, usually in seasonally or permanently wet sites.

S. bispinosa (Jacq.)W.Wight [29]

S. aculeata (Willd.)Poir. [29,32,47]; *Aeschynomene bispinosa* Jacq. [29].
Not climbing[29]; Herb or Shrub[29]; Perennial[29].
Indo-China: Cambodia(N)[29]; Laos(N)[29]; Thailand(N)[32]; Vietnam(N)[29]. Asia: India(N)[29]; Malaysia-ISO(N)[29]; Pakistan(N)[29].
Description[29,47].
Domestic[29]; Fibre[47].

S. cannabina (Retz.)Pers. [160]

S. cannabina var. *floribunda* Gagnep. [160].
Not climbing [29]; Herb or Shrub[29]; Perennial[29].
Indo-China: Vietnam [29].
Nguyen van Thuan et al. (1987: ref.29) treat this as a synonym of *S. sericea* but this is not in line with current practice.

S. grandiflora (L.)Poir. [29]

Robinia grandiflora L. [29].
Not climbing[29]; Tree[29]; Perennial[29].
Indo-China: Cambodia(I)[29]; Laos(I)[29]; Thailand(I)[32]; Vietnam(I)[29].
Description[29,47].
Environmental[47]; Food or Drink[29]; Medicine[29]; Wood[29].
Probably originates from Indonesia. Cultivated in all tropical areas [29].

S. javanica Miq. [29]

S. aculeata var. *paludosa* Bak. [29]; *S. grandiflora* sensu Miq. [29]; *S. paludosa* Prain [47]; *S. roxburghii* Merr. [32].
Not climbing[29]; Herb or Shrub[29]; Perennial[29].
Indo-China: Cambodia(N)[29]; Laos(N)[29]; Thailand(N)[29]; Vietnam(N)[29]. Asia: Burma(N)[29]; India(N)[29]; Indonesia-ISO(N)[29]; Philippines(N)[29] Taiwan(N)[29].
Description[29,47].
Environmental[29]; Food or Drink[29].

S. sericea (Willd.)Link [29]

S. aculeata (Willd.) Poir. var. *sericea* Bak. [29]; *S. polyphylla* Miq. [29]; *Coronilla sericea* Willd. [29].
Not climbing[29]; Shrub[29]; Perennial[29].
Indo-China: Cambodia(N)[29]; Laos(N)[47]; Thailand(N)[29]; Vietnam(N)[29] Asia: China(N) [29]; Indonesia-ISO(N)[29]; Sri Lanka(N)[29].
Description[29]; Illustration[29]
Domestic[47]; Fibre[47].

S. sesban (L.)Merr. [29]

S. aegyptiaca sensu auctt. [29,32,47]; *S. aegyptiaca* Poir. [29]; *S. aegyptiaca* var. *bicolor* Wight & Arn. [29]; *S. aegyptiaca* var. *concolor* Wight & Arn. [29]; *S. aegyptiaca* var. *picta* Prain [29]; *Aeschynomene sesban* L. [29].
Indo-China: Cambodia(N)[29]; Laos(N)[29]; Thailand(N)[32]; Vietnam(N)[29].
Not climbing[29]; Shrub[29]; Perennial[29].
Description[29,47].
Domestic[47]; Environmental[29]; Medicine[47].
Throughout tropical regions of the old world. Often cultivated [29].

SOPHOREAE

BOWRINGIA Benth.

A small genus of woody lianas with two species in West Africa, one in Madagascar and one in southeast Asia.

B. callicarpa Benth. [29]

Climbing[29]; Shrub[29]; Perennial[29].
Indo-China: Cambodia(N)[29]; Laos(N)[29]; Vietnam(N)[29]. Asia: China(N)[29]; Malaysia-ISO(N)[29]: Borneo[29].
Description[29]; Illustration[29].

ORMOSIA Jackson

A pantropical genus of trees or shrubs, with perhaps 100 species in South America and eastern Asia, Malesia and Australia. Some are ornamental; others yield timber. Five segregates have been recognized on the basis of dispersal mechanisms, but are not adopted here.

O. balansae Drake [29]

Macroule balansae (Drake)Yakovlev [29]; *O. elliptilimba* Merr. & Chun [29].
Not climbing[29]; Tree[29]; Perennial[29].
Indo-China: Vietnam(N)[29]. Asia: China(N)[29].
Description[29]; Illustration[29].

O. cambodiana Gagnep. [29]

Placolobium cambodianum (Gagnep.)Yakovlev [29].
Not climbing[29]; Tree[29]; Perennial[29].
Indo-China: Cambodia(N)[29].
Description[29]; Illustration[29].

O. chevalieri Niyomdham [29]

Not climbing[29]; Tree[29]; Perennial[29].
Indo-China: Cambodia(N)[29]; Vietnam(N)[29].
Description[29]; Illustration[29].

O. crassivalvis Gagnep. [29]

Placolobium crassivalve (Gagnep.)Yakovlev [29].
Not climbing[29]; Tree[29]; Perennial[29].
Indo-China: Vietnam(N)[29].
Description[29]; Illustration[29].
Endemic to Vietnam [29].

O. emarginata (Hook. & Arn.)Benth. [29]

O. glaberrima Y.C.Wu [29]; *Fedorovia emarginata* (Hook. & Arn.)Yakovlev [29].
Not climbing[29]; Tree[29]; Perennial[29].
Asia: China(N)[29]. Indo-China: Vietnam(N)[29].
Description[29].

O. fordiana Oliver [29]

Ruddia fordiana (Oliver)Yakovlev [29].
Not climbing[29]; Tree[29]; Perennial[29].
Indo-China: Thailand(N)[29]; Vietnam(N)[29]. Asia: Burma(N)[29]; China(N)[29].
Description[29,76]; Illustration[29,76].
Medicine[29].

O. grandistipulata Whitm. [76]

Not climbing[76]; Tree[76]; Perennial[76].
Indo-China: Thailand(N)[76]. Asia: Malaysia-ISO(N)[76].
Description[76]; Illustration[76].

O. henryi Prain [29]

O. henryi Hemsley & Wilson [29]; *O. mollis* Dunn [29]; *Fedorovia henryi* (Prain)Yakovlev [29].
Not climbing[29]; Tree[29]; Perennial[29].
Indo-China: Thailand(N)[29]; Vietnam(N)[29]. Asia: China(N)[29].
Description[29,76]; Illustration[29,76].

O. hoaensis Gagnep. [29]

Placolobium hoaense (Gagnep.)Yakovlev [29].
Not climbing[29]; Tree[29]; Perennial[29].
Indo-China: Vietnam(N)[29].
Description[29]; Illustration[29].
Endemic to Vietnam [29].

O. inflata Merr. & L.Chen [29]

O. merrilliana L.Chen [29]; *Trichocyamos inflatum* (Merr. & Chun)Yakovlev [29]; *Trichocyamos merrillianum* (L.Chen)Yakovlev [29].
Not climbing; Tree[29]; Perennial[29].
Indo-China: Vietnam(N)[29]. Asia: China(N)[29].
Description[29]; Illustration[29].

O. kerrii Niyomdham [76]

Not climbing[76]; Tree[76]; Perennial[76].
Indo-China: Thailand(N)[76]. Asia: Malaysia-ISO(N)[76].
Description[76]; Illustration[76].

O. laosensis Niyomdham [29]

Not climbing[29]; Tree[29]; Perennial[29].
Indo-China: Laos(N)[29]; Vietnam(N)[29].
Description[29]; Illustration[29].

O. macrodisca Bak. [76]

O. basilanensis Merr. [76]; *O. clemensii* Merr. [76]; *O. grandifolia* Merr. [76]; *O. monchyana* Koord. [76]; *O. palembanica* S.Moore [76]; *Placolobium sumatranum* Miq. [76].
Not climbing[76]; Tree[76]; Perennial[76].
Indo-China: Thailand(N)[76]. Asia: Indonesia-ISO(N)[76]; Malaysia-ISO(N)[76], Borneo[76]; Philippines(N)[76].
Description[76]; Illustration[76].

O. pinnata (Lour.)Merr. [29]

O. hainanensis Gagnep. [29]; *Cynometra pinnata* Lour. [29]; *Fedorovia pinnata* (Lour.)Yakovlev [29].
Not climbing[29]; Tree[29]; Perennial[29].
Indo-China: Thailand(N)[29]; Vietnam(N)[29]. Asia: China(N)[29].
Description[29,76]; Illustration[29,76].
Wood[29].

O. poilanei Niyomdham [29]
Not climbing[29]; Tree[29]; Perennial[29].
Indo-China: Vietnam(N)[29].
Description[29]; Illustration[29].
Endemic to cental Vietnam [29].

O. robusta Bak. [76]
Arillaria robusta Kurz [76]; *Placolobium robustum* (Roxb.)Yakovlev [76].
Not climbing[76]; Tree[76]; Perennial[76].
Indo-China: Thailand(N)[76]. Asia: Burma(N)[76]; India(N)[76].
Description[76].

O. simplicifolia Merr. & L.Chen [29]
Fedorovia simplicifolia (Merr. & L.Chen)Yakovlev [29].
Not climbing[29]; Tree[29]; Perennial[29].
Indo-China: Vietnam(N)[29]. Asia: China(N)[29].
Description[29]; Illustration[29].
Endemic to S.China & N.Vietnam [29].

O. sumatrana (Miq.)Prain [29]
O. coarctata sensu auct. [29]; *O. decemjuga* (Miq.)Prain [29]; *O. euphorioides* Gagnep. [29]; *O. microsperma* Bak. [29]; *O. septemjuga* (Miq.)Prain [29]; *O. yunnanensis* Prain [29]; *Chaenolobium decemjugum* Miq. [29]; *Chaenolobium septemjugum* Miq. [29]; *Macrotropis sumatrana* Miq. [29].
Not climbing[29]; Tree[29]; Perennial[29].
Indo-China: Laos(N)[29]; Thailand(N)[29]; Vietnam(N)[29]. Asia: China(N)[29]; Indonesia-ISO(N)[29]; Malaysia-ISO(N)[76].
Description[29,76].

O. tavoyana Prain [4,76,77]
Placolobium tavoyanum (Prain)Yakovlev [78].
Not climbing[4]; Tree[4]; Perennial[4].
Indo-China: Thailand(N)[4]. Asia: Burma(N)[76].
Description[77].
Kerr 16680: 'tree'; *Kerr* 18499: 'Thailand' [4].

O. tonkinensis Gagnep. [29]
Not climbing[29]; Tree[29]; Perennial[29].
Indo-China: Vietnam(N)[29].
Description[29].
Only known from the type (*Balansa* 2253) [29].

O. tsangii L.Chen [29]
Fedorovia tsangii (L.Chen)Yakovlev [29].
Not climbing[29]; Tree[29]; Perennial[29].
Indo-China: Vietnam(N)[29].
Description[29]; Illustration[29].
Endemic to Vietnam[29].

O. xylocarpa Merr. & L.Chen [29]
Fedorovia xylocarpa U(Merr. & L.Chen)Yakovlev [29]; *O. polysperma* L.Chen [29].
Not climbing[29]; Tree[29]; Perennial[29].
Indo-China: Laos(N)[29]; Vietnam(N)[29]. Asia: China(N)[29].
Description[29]; Illustration[29].

PLACOLOBIUM Miq.

One of the segregates of *Ormosia* (q.v.).

P. ellipticum N.D.Khoi & Yakovlev [29]
Not climbing[29]; Tree[29]; Perennial[29].
Indo-China: Vietnam(N)[29].
Description[29]; Illustration[79].
Probably *Ormosia* as treated by Nguyen van Thuan, but no material seen by him; no combination exists in *Ormosia* [29,79].

P. vietnamense N.D.Khoi & Yakovlev [79]

Not climbing[29]; Tree[29]; Perennial[29].
Indo-China: Vietnam(N)[29].
Description[29]; Illustration[79].
Probably *Ormosia* as treated by Nguyen van Thuan et al. but they saw no material[79]; no combination exists in *Ormosia* [29,29].

SOPHORA L.

A genus of about 50 species of trees and shrubs, widespread, but mainly in the warm temperate zones. A former subgenus has recently been recognized at generic level (*Styphnolobium* - q.v.).

S. dunnii Craib [173]

S. dispar Craib [173]
Not climbing[173]; Shrub[173]; Perennial[173].
Indo-China: Thailand(N)[173]. Asia: China(N)[173].
Description [173].
Nguyen van Thuan et al. [29] regarded *S. dispar* as a synonym of *S. glauca* (q.v.).

S. exigua Craib [29]

S. violacea var. *pilosa* Gagnep. [29]; *S. violacea* subsp. *pilosa* (Gagnep.)Yakovlev [29].
Not climbing[29]; Shrub or Herb[29]; Perennial[29].
Indo-China: Cambodia(N)[29]; Thailand(N)[29].
Description[29,76]; Illustration[76].

S. glauca DC. [29]

S. velutina Lindley [29,76].
Not climbing[29]; Shrub[29]; Perennial[29].
Indo-China: Thailand(N)[76]; Vietnam(N)[29]. Asia: India(N)[29]; Malaysia-ISO(N)[76]; Philippines(N)[76].
Description[29,76]; Illustration[29].
Tsoong Pu-chiu & Ma Chi-yun (1981) [173] regard *S. velutina* as a good species native to China.

S. prazeri Prain [173]

S. bhutanica Ohashi [76]; *S. duclouxii* Gagnep. [76]; *S. mairei* Pampan. [76]; *S. wightii* sensu auct. [173]; *S. wilsonii* Craib [76]; *Millettia esquirolii* Leveille [76].
Not climbing[76]; Shrub or Tree[76]; Perennial[76].
Indo-China: Thailand(N)[76]. Asia: China(N)[76]; India(N)[76]; Indonesia-ISO(N)[76].
Description[76]; Illustration[76].
Tsoong & Ma (1981) regard *S. mairei* and *S. wilsonii* Craib as synonyms of a separate variety, var. *mairei* (Pamp.)Tsoong.

S. tomentosa L. [29]

Not climbing[29]; Shrub[29]; Perennial[29].
Indo-China: Cambodia(N)[29]; Thailand(N)[76]; Vietnam(N)[29].
Description[29,76]; Illustration[29].
Medicine[29].
Widespread in all tropical & subtropical regions [29].

S. tonkinensis Gagnep. [29]

S. subprostrata Chun & T.Chen [29]; *Cephalostigmaton tonkinense* (Gagnep.)Yakovlev [29].
Not climbing[29]; Shrub[29]; Perennial[29].
Indo-China: Vietnam(N)[29]. Asia: China(N)[29].
Description[29,47]; Illustration[29,47].

STYPHNOLOBIUM Schott

A genus of nine species, all trees, mainly in central and north America but with one species in eastern Asia; this is widely planted elsewhere. Differs from *Sophora* in its habit, presence of stipels, and indehiscent fruits. See Sousa and Rudd, Ann. Missouri Bot. Gard. 80: 270–283.

S. japonicum (L.) Schott
Sophora japonica L.
Not climbing[29]; Tree[29]; Perennial[29].
Indo-China: Vietnam(I)[29]. Asia: China(N)[29]; Japan(N)[29].
Description[29]; Illustration[29].
Chemical Products[29]; Environmental[29].
Originally from Japan & China. Often cultivated [29].

TRIFOLIEAE

MEDICAGO L.

A genus of small, mostly annual herbs with coiled pods, mainly from the Mediterranean and Irano-Turanian regions. Widely introduced, sometimes as fodder crops but also accidentally through their hooked fruits.

M. polymorpha L. [29]
M. denticulata Willd. [29]; *M. hispida* Gaertn. [167].
Not climbing[29]; Herb[29]; Perennial[29].
Indo-China: Vietnam(I) [29].
Description[29]; Illustration[29].

MELILOTUS L.

Tall annual herbs, mainly from southern Europe and western Asia. About 20 species, some widely introduced either accidentally or as fodder crops.

M. indicus (L.)All. [32]
M. parviflora Desf. [32]; *Trifolium indicum* L. [167].
Indo-China: Thailand(N) [32].
Not climbing[29]; Herb[29]; Annual[29].
This record may refer to *M. suaveolens*.

M. suaveolens Ledeb. [29]
Not climbing[29]; Herb[29]; Annual or Perennial[29].
Indo-China: Laos(N)[29]; Vietnam(N)[29]. Asia: China(N)[29]; India(N)[29]; Japan(N)[29]; Taiwan(N)[29].
Description[29]; Illustration[29].
Medicine[29].

PAROCHETUS D.Don

Small creeping herbs from the mountains of tropical Asia and Africa. The plants from the two continents have recently been shown to be different species.

P. communis D.Don [29]
P. maculata R.Br. [29]; *P. major* D.Don [29].
Not climbing[29]; Herb[29]; Perennial.
Indo-China: Thailand(N)[4]; Vietnam(N)[29]. Asia: Burma(N)[29]; India(N)[29]; Indonesia-ISO(N)[29]; Sri Lanka(N)[29].
Description[29]; Distribution Map[49]; Illustration[29]
Garrett 91: 'Thailand' [4].

VICIEAE

LATHYRUS L.

A genus of about 150 species, mostly climbing herbs. Most diverse in Europe and the Mediterranean region, but widespread in the north temperate zone and extending into the tropics on mountains. Some are cultivated as fodder and for their ornamental flowers.

Lathyrus palustris L. [29]

Not climbing[29]; Herb[29]; Perennial[29].
Indo-China: Laos(N)[29]. Asia: China(N)[29]; India(N)[29]; Japan(N)[29].
Description[29,47]; Illustration[29,47].

LENS Miller

Two or three species of annual herbs from the Mediterranean region, one widely cultivated for its edible seeds (lentils).

L. culinaris Medikus [29]

L. esculenta Moench [29]; *Ervum lens* L. [29].
Not climbing[29]; Herb[29]; Perennial[29].
Indo-China: Vietnam(I)[29].
Description[29]; Illustration[29].
Originating from Southern Europe, this species is cultivated in Africa & Asia [29].

PISUM L.

Climbing herbs from the Mediterranean and Irano-Turanian regions, widely cultivated for their edible seeds and pods. Number of species disputed but probably only two; some systems recognize a single species with three subspecies.

P. arvense L. [29]

P. sativum L. subsp. *arvense* (L.)Asch. & Graebner [29].
Climbing[29]; Herb[29]; Perennial[29].
Indo-China: Laos(I)[29]; Vietnam(I)[29].
Description[29]; Illustration[29].
Species cultivated throughout Europe & Asia [29].

P. sativum L. [29]

Climbing[29]; Herb[29]; Perennial[29].
Indo-China: Vietnam(I)[29].
Description[29]; Illustration[29].
Food or Drink[29].
Cultivated extensively in Europe & Asia [29].

VICIA L.

About 150 species, mainly climbing herbs of the north temperate regions, extending into upland tropical areas. Some species, especially *V. faba*, are important food crops; others are grown as fodder.

V. faba L. [29]

Faba vulgaris Moench [29].
Not climbing[29]; Herb[29]; Annual[29].
Indo-China: Vietnam(I)[29].
Description[29].
Food or Drink[29].
Probably originates from West Asia. Cultivated [29].

V. peregrina L. [29]
Climbing[29]; Herb[29]; Annual[29].
Indo-China: Vietnam(N)[29].
Description[29].
Cultivated in Europe & Asia [29].

V. sativa L. [29]
Climbing or not[29]; Herb[29]; Annual[29].
Indo-China: Vietnam(U)[29].
Description[29]; Illustration[29].
Origin unknown;widely distributed;often an escape from cultivation; occurs in all temperate regions of the world up to 1200 m [29].

V. tenuifolia Roth [29]
Climbing[29]; Herb[29]; Annual[29].
Indo-China: Vietnam(U)[29].
Description[29].
Weedy species;temperate regions of Europe and the NE Himalayas [29].

Note added in proof:

CLITORIA
The varieties of *C. hanceana* and *C. macrophylla* mentioned on pp. 101 and 102 have now been formally described by Fantz, P.R. (1993). Notes on *Clitoria* (Leguminosae) in Southeast Asia. Novon 3: 352–355.

He recognises the following:

C. hanceana var. **hanceana**, var. **laureola** Gagnep., var. **latifolia** Fantz, var. **petiolata** Fantz and var. **thailandica** Fantz.
C. macrophylla var. **macrophylla**, var. **sericea** Fantz and var. **stipulacea** Fantz.

BIBLIOGRAPHY

1. Nielsen, I. (1985). Leguminosae - Mimosoideae. In: Flora of Thailand, Vol. 4 Pt.2. Bangkok: The Forest Herbarium.
2. Nielsen, I. (1981). 19. Légumineuses - Mimosoidées. In: Flore du Cambodge du Laos et du Viêt-nam. Paris: Museum National d'Histoire Naturelle.
3. Polhill, R.M. (1990). 80. Légumineuses. Flore des Mascareignes.Mauritius: Sugar Industry Research Institute; Paris: ORSTOM; Kew: Royal Botanic Gardens.
4. Heald, J. (1992). Specimen in Herb., Royal Botanic Gardens, Kew.
5. Nielsen, I. (1980). Notes on Indo-Chinese Mimosaceae. Adansonia, sér. 2, 19: 339-363.
6. Nielsen, I. (1985). The Malesian species of *Acacia* and *Albizia* (Leguminosae - Mimosoideae). Opera Botanica 81: 1-50.
7. Ridley, H.N. (1922). The Flora of the Malay Peninsula. Vol. 1: Polypetalae. London: L. Reeve & Co.
8. Nielsen, I. (1979). Notes on the genera *Archidendron* F. Muell. and *Pithecellobium* Martius in mainland SE Asia. Adansonia, ser. 2, 19: 3–37.
9. Nielsen, I. (1979). Notes on the genus *Albizia* Durazz. (Leguminosae - Mimosoideae) in mainland SE Asia. Adansonia, ser. 2, 19: 199–229.
10. Nielsen, I., Baretta-Kuipers, T. & Guinet, P. (1984). The genus *Archidendron* (Leguminosae - Mimosoideae). Opera Botanica 76: 1–120.
11. Oliver, D. (1891). *Pithecolobium balansae.* Hooker's Ic. Pl. 20: t.1976.
12. Merrill, E.D. (1928). An enumeration of Hainan Plants. Lingnan Sci. Journ. 5: 1–186.
13. Kostermans, A.J.G.H. (1960). Miscellaneous Botanical Notes. 1. Reinwardtia 5: 233–254.
14. Kostermans, A.J.G.H. (1966). Notes on some Asian Mimosaceous genera. Adansonia ser. 2, 6: 351–373.
15. Brenan, J. & Brummitt, R. (1965). The variation of *Dicrostachys cinerea* (L.) Wight & Arn. Bol. Soc. Brot., ser. 2, 39: 61–115.
16. Larsen, K., Larsen, S.S & Vidal, J.E. (1980). 18. Légumineuses - Césalpinioidées. In: Flore du Cambodge du Laos et du Viêt-nam. Paris: Museum National d'Histoire Naturelle.
17. Larsen, K., Larsen, S.S. & Vidal, J.E. (1984). Leguminosae - Caesalpinioideae. In: Flora of Thailand, Vol. 4 Pt.1. Bangkok: The Forest Herbarium.
18. Hattink, T.A. (1974). A revision of Malesian *Caesalpinia* including *Mezoneuron* (Leguminosae - Caesalpiniaceae). Reinwardtia 9: 1–69
19. Gagnepain, F. (1916). In Lecomte, H. (Ed.): Flore Générale de L'Indochine 2, 3: Légumineuses: Caesalpiniées (fin), Papilionées.
20. Larsen, K. (1989). *Dialium patens* Bak. (Leguminosae - Caesalpinioideae) new to Thailand. Thai Forest Bulletin (Botany) 18: 82–83.
21. Knaap van Meeuwen, M.S. (1970). A revision of four genera of the tribe Leguminosae - Caesalpinioideae - Cynometreae in Indomalesia and the Pacific. Blumea 18: 1–52.
22. De Wit, H.C.D. (1949). Revision of the genus *Sindora* Miquel (Leguminosae). Bull. Jard. Bot. Buitenzorg ser. 3, 18: 5–82.
23. Zuijderhoudt, G.F.P. (1968). A revision of the genus *Saraca* L. (Leguminosae - Caesalipinioideae). Blumea 15: 413–425.
24. De Wit, H.C.D. (1956). A revision of Malaysian Bauhinieae. Reinwardtia 3: 381–539.
25. Larsen, K. & Larsen, S.S. (1980). Notes on the genus *Bauhinia* in Thailand. Thai Forest Bulletin (Bot.) 13: 37–46.
26. Lock, J.M. (1988). *Cassia* sens. lat. (Leguminosae - Caesalpinioideae) in Africa. Kew Bull. 43: 333–342.

27. Irwin, H.S. & Barneby R.C. (1982). The American Cassiinae, a synoptical revision of Leguminosae Tribe Cassieae Subtribe Cassiinae in the New World. Mem. New York Bot. Gard. 35: 1–454.
28. Chun Woon-Young. (1946). A new genus in the Chinese flora [*Zenia*]. Sunyatsenia 6: 195–198
29. Nguyen van Thuan, Dy Phon, P. & Niyomdham, C. (1987). 23: Légumineuses – Papilionoidées. In: Flore du Cambodge du Laos et du Viêtnam. Paris: Museum Nationale d'Histoire naturelle.
30. Gagnepain, F. (1920). In Lecomte, H. (Ed.): Flore Générale de L'Indochine 2, 5: Légumineuses: Papilionacées (fin).
31. Ohashi, H. (1973). The Asiatic species of *Desmodium* and its allied genera (Leguminosae). Ginkgoana 1.
32. Craib, W.G. (1928). Florae Siamensis Enumeratio 1(3). Connaraceae & Leguminosae. Siam Society, Bangkok.
33. Ohashi, H. (1971). A monograph of the subgenus *Dollinera* of the genus *Desmodium* (Leguminosae). In: Hara, H. (ed.) Flora of Eastern Himalaya, Second Report. Univ. Mus. Univ. Tokyo, Bulletin 2: 259-320.
34. Ohashi, H. (1990). *Desmodium schubertiae* (Leguminosae). A new species from Cambodia and Vietnam. J. Arn. Arbor. 71: 381–384.
35. De Candolle, A.P. (1825). Notice sur quelques genres et espèces nouvelles de Légumineuses. Ann. Sci. Nat. 4: 90–103.
36. Bresser, M. (1978). Monograph of the genus *Phylacium* (Leguminosae). Blumea 24: 485–493
37. Pramanik, A. & Thothathri K. (1983). Notes on the taxonomy, distribution and ecology of *Lespedeza juncea* complex with special reference to India. J. Jap. Bot. 58: 331–337.
38. Huang, T.C. & Ohashi, H. (1977). Leguminosae. In: Flora of Taiwan 3. Epoch Publishing Co., Ltd., Taipei, Taiwan.
39. Schindler, A.K. (1916). Desmodiinae Novae. Bot. Jahrb. Syst.54: 51–68.
40. Schindler, A.K. (1912). Das Genus *Campylotropis*. Feddes Rep. Nov. Sp. 11: 338--347.
41. Kurz, S. (1874). New Burmese Plants (III). J. Asiat. Soc. Bengal 42: 227–254.
42. Schindler, A.K. (1926). Desmodii generumque affinum species et combinationes novae II. Feddes Rep. Nov. Sp. 22: 250–288.
43. Ohashi, H. (1971). A taxonomic study of the tribe Coronilleae (Leguminosae) with a special reference to pollen morphology. Journ. Fac. Sci. Univ. Tokyo (Botany) 11: 25–92.
44. Miquel, F.A.W. et al. (1852). Plantae Junghuhnianae, part 2. Leiden.
45. De Haas, A.J.P., Bosman, M.T.M. & Geesink, R. (1980). *Urariopsis* reduced to *Uraria* (Leguminosae – Papilionoideae). Blumea 26: 439–444.
46. Craib, W.G. (1912). Contributions to the Flora of Siam. Additamenta. Bull. Misc. Inform., Kew 1912: 144–155.
47. Gagnepain, F. (1913). In Lecomte, H. (Ed.): Flore Générale de L'Indochine 2, 2. Légumineuses: Mimosées et Caesalpiniées.
48. Niyomdham, C. (1978). A revision of the genus *Crotalaria* Linn. (Papilionaceae) in Thailand. Thai Forest Bulletin (Bot.) 11: 105–181.
49. Dy Phon, P. (1981). Taxons nouveaux pour la flore de l'Indochine. Bull. Mus. Nat. Hist. Nat. (Paris) sect. B, Adansonia 3:111–119.
50. Rudd, V.E. (1959). The genus *Aeschynomene* (Leguminosae – Papilionoideae) in Malaysia. Reinwardtia 5: 23–36.
51. Niyomdham, C. (1989). A revision of the genus *Euchresta* Bennett (Leguminosae – Papilionoideae) in Thailand. Thai Forest Bulletin (Bot.) 18: 80–81.

52. De Kort, I. & Thijsse G. (1984). A revision of the genus *Indigofera* (Leguminosae – Papilionoideae) in Southeast Asia. Blumea 30: 89–151.
53. Pyramarn, K. (1986). *Indigofera kasinii* sp. nov. (Papilionaceae) in Thailand. Thai Forest Bulletin (Bot.) 16: 207–210.
54. DuPuy, D.J., Labat, J.-N. & Schrire, B.D. (1993). The separation of two previously confused species in the *Indigofera spicata* complex (Leguminosae: Papilionoideae. Kew Bulletin 48: 727–733.
55. Loureiro, J. de (1790). Flora Cochinchinensis, Vols. 1–2. Lisbon
56. Merrill, E.D. (1928). The identity of the genus *Sarcodum* Louriero. J. Bot. 66: 264–265.
57. Polhill, R.M. (1971). Some observations on generic limits in Dalbergieae – Lonchocarpineae Benth. (Leguminosae). Kew Bull. 25: 258–274.
58. Gagnepain, F. (1915). Papilionacées nouvelles ou critiques. Notul. Syst. (Paris) 3(6):180–192.
59. Burtt, B.L. & Chermsirivathana, C. (1971). A second species of *Afgekia* (Leguminosae). Notes Roy. Bot. Gard. Edinburgh 31:131–133.
60. Craib, W.G. (1927). Contributions to the Flora of Siam. Additamentum XXIII. Bull. Misc. Inform., Kew 1927: 374–394.
61. Geesink, R. (1984). Scala Millettiearum. A survey of the genera of the Millettieae (Leguminosae – Papilionoideae) with methodological considcrations. Leiden Botanical Series 8. Leiden University Press.
62. Dunn, S.T. (1911). *Adinobotrys* and *Padbruggea*. Bull. Misc. Inform., Kew 1911: 193–198.
63. Craib, W.G. (1927). Contributions to the Flora of Siam. Additamentum XX. Bull. Misc. Inform., Kew 1927: 56–72.
64. Buijsen, J.R.M. (1988). Revision of the genus *Fordia* (Papilionaceae: Millettieae). Blumea 33: 239–261.
65. Merrill, E.D. (1934). New Sumatran Plants. 1. Papers Mich. Acad. Sci. 19: 149–204
66. Whitmore, T.C. (1972). Tree Flora of Malaya, Vol. 1. A Manual for Foresters. London: Longman.
67. Burkill, I.H. (1935). Some changes in plant names. Bull. Misc. Inform., Kew 1935: 316–319.
68. Prain, D. in King, G. (1897). Materials for a Flora of the Malayan Peninsular. Journ.As.Soc.Beng. 66,2: 1–345.
69. Thothathri, K. (1982). Leguminosae: genus *Derris*. Fascicles of Flora of India, Fascicle 8. Botanical Survey of India.
70. Dunn, S.T. (1912). A revision of the genus *Millettia* Wight & Arn. J. Linn. Soc.(Bot.) 41: 123–243.
71. Miquel, F.A.W. (1855). Flora Indiae Batavae (Flora van Nederlandsch Indie). Pts. 1–4. Amsterdam: C.G. van der Post.
72. Kurz, S. (1873). New Burmese Plants (II). J. Asiat. Soc. Bengal 42: 59– 110.
73. Bosman, M.T.M. & de Haas A.J.P. (1983). A revision of the genus *Tephrosia* (Leguminosae – Papilionoideae) in Malesia. Blumea 28: 421–487.
74. Stirton, C.H. (1981). Studies in the Leguminosae – Papilionoideae of Southern Africa. Bothalia 13: 317–325.
75. Lock, J.M. & Simpson, K. (1991). Legumes of West Asia: a Check-List. Royal Botanic Gardens, Kew
76. Niyomdham, C. (1980). Preliminary revision of Tribe Sophoreae (Leguminosae – Faboideae) in Thailand: *Ormosia* Jacks. and *Sophora* Linn. Thai Forest Bulletin (Bot.) 13: 1–22.
77. Prain, D. (1904). The Asiatic species of *Ormosia*. J. Asiat. Soc. Bengal 73: 45–46.

78. Yakovlev, G.P. (1973). De Generibus *Sweetia* Spreng., *Machaerium* Pers., *Angylocalyx* Taub., *Federovia* Yakovl., *Placolobium* Miq., et *Ormosia* Jacks. (Fabaceae). Notulae Systematicae. Novosti Sist. Vyssh. Rast. 10: 190–196.
79. Khoi, N.D. & Yakovlev, G.P. (1981). Fabaceae in Viet-nam. 1. *Placolobium* , *Sophora*. Bot. Zhurn. 66: 1770–1771
80. Prain, D. (1904). The species of *Dalbergia* of South-eastern Asia. Annals Royal Bot. Gard. Calcutta 10: 1–114.
81. Hance, H.F. (1882). Spicilega Florae Sinensis: Diagnoses of new, and habitats of rare or hitherto unrecorded Chinese plants – VI. J. Bot. 20: 2–6.
82. Thothathri, K. (1987). Taxonomic revision of the Dalbergieae in the Indian Subcontinent. Calcutta: Botanical Survey of India.
83. van der Maesen, L.J.G. (1985). Revision of the genus *Pueraria* DC. with some notes on *Teyleria* Backer (Leguminosae). Agricultural Univ. Wageningen Papers 85–1
84. Rojo, J.P. (1972). *Pterocarpus* (Leguminosae – Papilionoideae) revised for the World. Phanerogamarum Monographiae, Tomus 5. pp.119.
85. Merrill, E.D. (1910).An enumeration of Philippine Leguminosae, with keys to the genera and species (concluded). Philipp. J. Sci. 5(2): 95–136.
86. Nguyen van Thuan (1979). 17. Légumineuses – Phaseolées. In: Flore du Cambodge du Laos et du Viêtnam. Paris: Museum Nationale d'Histoire naturelle.
87. van der Maesen, L.J.G. (1985). *Cajanus* DC. & *Atylosia* Wight & Arn. (Leguminosae). Agricultural Univ. Wageningen Papers 85–4.
88. Bentham, G. (1860). A synopsis of the Dalbergieae, a tribe of the Leguminosae. Journ. Linn. Soc. 4, Suppl.: 1–128.
89. Craib W.G. (1924). Contributions to the Flora of Siam. Additamentum XIV. Bull. Misc. Inform., Kew 1924: 82–98.
90. Craib, W.G. (1926). Contributions to the Flora of Siam. Additamentum XVIII. Bull. Misc. Inform., Kew 1926: 154–174.
91. Backer, C.A. (1939). Notes on the Flora of Java. 1. *Teyleria*, a new genus of Leguminosae. Bull. Jard. Bot. Buitenzorg ser. 3, 16: 107–109.
92. van der Maesen, L.J.G. (1988). Miscellaneous Notes: No.37. Two corrections to the nomenclature in the revision of *Pueraria* DC. J. Bombay Nat. Hist. Soc. 85: 233–234.
93. Nielsen, I. & Guinet P. (1992). Synopsis of *Adenanthera* (Leguminosae – Mimosoideae). Nordic J. Bot. 12: 85–114
94. Dunn, S.T. (1903). Descriptions of new Chinese plants. Journ. Linn. Soc., Bot. 35: 491
95. Khoi, N.D. & Yakovlev, G.P. (1982). The supplements to the flora of the Fabaceae from Vietnam: 1. [in Russian]. Bot. Zhurn. 67: 1540–1543.
96. Labat, J.-N. (1991). *Abrus longibracteatus*, une espèce nouvelle de Leguminosae – Papilionoideae du Laos et du Vietnam. Bull. Mus. natl Hist. nat., Paris, 4e sér. 13, sect. B, Adansonia: 167–71.
97. Ohashi, H. (1966). Leguminosae. In Hara,H. (Ed.) Flora of Eastern Himalaya. Pp. 135–166. Tokyo: University of Tokyo Press.
98. Ohashi, H., Tateishi, Y., Murata, G. & Kato, M. (1982). Contributions towards the plant taxonomy and distribution in the Himalayan elements. J. Jap. Bot. 57: 1–9.
99. Grierson, P. & Long, D. (1987). Flora of Bhutan, Vol.1 Pt.3. Edinburgh: Royal Botanic Garden.
100. Lock, J.M. (1993). Notes and specimens in Herb. Royal Botanic Gardens, Kew.

101. Predeep, S.V. & Nayar, M.P. (1990). Notes on *Shuteria ferruginea* (Benth.) Baker (Leguminosae - Papilionoideae). J. Jap. Bot.65: 369–373.
102. Sanjappa, M. (1989). Revision of the genera *Butea* Roxb. ex Willd. and *Meizotropis* Voight (Fabaceae). Bull. Bot. Surv. India 29: 199–225.
103. Wei, Y-T & Lee, S-K. (1985). New material for Chinese Leguminosae. Guihaia 5: 157–174.
104. Sanjappa, M. (1992). Legumes of India. Dehra Dun: Bishen Singh Mahendra Pal Singh.
105. Khoi, N.D. & Yakovlev, G.P. (1982). The supplements to the flora of the Fabaceae from Vietnam: 1. [in Russian]. Bot. Zhurn. 67: 1540–1543.
106. Merrill, E.D. (1935). A commentary on Loureiro's 'Flora Cochinchinensis'. Trans. Amer. Phil. Soc. 24: 1–445.
107. Fantz, P.R. (1979). A new species of *Clitoria* subgenus *Neurocarpus* (Leguminosae) and a new species endemic to Thailand. Brittonia 31: 115–118.
108. Niyomdham, C. (1992). Notes on Thai and Indo-Chinese Phaseoleae (Leguminosae - Papilionoideae). Nordic J. Bot. 12: 339–346.
109. King, G. & Prain, D. (1898). Descriptions of some new plants from the North-Eastern Frontiers of India. J. Asiat. Soc. Bengal 67: 284–305.
110. King, G., Duthie, J.F. & Prain, D. (1906). Descriptions of a second century of new and rare Indian plants. Ann. Roy. Bot. Gard., Calcutta 9: 1–80.
111. Verdcourt, B. (1979). A Manual of New Guinea Legumes. Lae: Office of Forests, Division of Botany.
112. Kerr, A.F.G. (1941). Contributions to the flora of Thailand, Additamentum LIV. Bull. Misc. Inform, Kew 1941: 8–21.
113. Verdcourt, B. (1970). Studies in the Leguminosae - Papilionoideae for the Flora of Tropical East Africa: III. Kew Bulletin 24: 379–447.
114. Prain, D. (1897). Noviciae Indicae XV. Some additional Leguminosae. J. Asiat. Soc. Bengal 66: 347–518.
115. Nguyen van Thuan (1977). Phaseolées Asiatiques Nouvelles. Adansonia, Sér 2, 16: 509–514.
116. Wu, Y.C. (1940). Beiträge zur Kenntnis der Flora Süd-Chinas. Bot. Jahrb. Syst. 71: 169–199.
117. Van Welzen, P.C. & den Hengst, S. (1985). A revision of the genus *Dysolobium* (Papilionaceae) and the transfer of subgenus *Dolichovigna* to *Vigna*. Blumea 30: 363–383.
118. Maréchal, R. Mascherpa, J-M., & Stanier, F. (1978). Etude taxonomique d'un groupe complexe d'espèces des genres *Phaseolus* et *Vigna* (Papilionaceae) sur la base des données morphologiques et polliniques, traitées par l'analyse informatique. Boissiera 28: 1–273.
119. Krukoff, B.A. & Barneby, R.C. (1974). Conspectus of species of the genus *Erythrina*. Lloydia 37: 332–459.
120. Craib, W.G. (1927). Contributions to the Flora of Siam. Additamentum XX. Bull. Misc. Inform., Kew 1927: 56–72.
121. Van Meeuwen, M.S. Nooteboom, H.P. & van Steenis, C.G.G.J. (1960). Preliminary revisions of some genera of Malaysian Papilionaceae. 1. Reinwardtia 5: 419–456.
122. Craib, W.G. (1911). List of Siamese plants with descriptions of new species. Bull. Misc. Inform., Kew 1911: 7–60.
123. Wilmot-Dear, C.M. (1992). A revision of *Mucuna* in Thailand, Indochina & the Malay Peninsular. Kew Bull. 47: 203–245.
124. Wilmot-Dear, C.M. (1984). A revision of *Mucuna* (Leguminosae - Papilionoideae) in China & Japan. Kew Bull. 39: 23–65.

125. Wilmot-Dear, C.M. (1990). A revision of *Mucuna* (Leguminosae – Papilionoideae) in the Pacific. Kew Bull. 45: 1–35.
126. Wilmot-Dear, C.M. (1991). *Mucuna hainanensis* Hayata subsp. *multilamellata* Wilmot-Dear, a new name for a long-known taxon (Leguminosae – Phaseoleae) and a key to related species. Kew Bull. 46: 206–212.
127. Wilmot-Dear, C.M. (1987). A revision of *Mucuna* (Leguminosae – Papilionoideae) in the Indian Subcontinent & Burma. Kew Bull. 42: 23–46.
128. Wilmot-Dear, C.M. (1993). A new species of *Mucuna* (Leguminosae – Phaseoleae) from Thailand, and a revised key to the species in Thailand, Indochina and the Malay Peninsular. Kew Bull. 48: 29–35.
129. Sørensen, M. (1988). A taxonomic revision of the genus Pachyrhizus (Fabaceae – Phaseoleae). Nordic J. Bot. 8: 167–192.
130. Jackson, B.D. (1893). Index Kewensis. Oxford: Clarendon Press.
131. Ohashi, H. & Tateishi, Y. (1978). A new subspecies of *Rhynchosia harae* from Laos and Thailand. J. Jap. Bot. 53: 142–144.
132. Ohashi, H. & Tateishi, Y. (1977). The genus *Rhynchosia* in Nepal. Bot. Mag. Tokyo 90: 219–233.
133. Nguyen Van Thuan (1972). Révision du genre *Shuteria* (Papilionaceae). Adansonia, sér. 2, 12: 291–305.
134. Ohashi, H. (1988). New names for two Asiatic legumes. J. Jap. Bot. 63: 159–160.
135. Ridder-Numan, J.W.A. & Wiriadinata, H. (1985). A revision of the genus *Spatholobus* (Leguminosae – Papilionoideae). Reinwardtia 10: 139–205.
136. Ridder-Numan, J.W.A. (1992). *Spatholobus* (Leguminosae – Papilionoideae): a new species and some taxonomic notes. Blumea 37: 63–71.
137. Babu, C.R., Sharma, S.K. & Johri, B.M. (1987). Leguminosae – Papilionoideae: tribe Phaseoleae. Bull. Bot. Surv. India 27: 1–28.
138. Verdcourt, B. (1970). Studies in the Leguminosae – Papilionoideae for the Flora of Tropical East Africa. 1. Kew Bull. 24: 1–70.
139. Ali, S.I. (1963). A taxonomic study of the genus *Psoralea* from West Pakistan. Biologia (Lahore) 9: 17–22.
140. Anon, 1803. Curtis's Botanical Magazine 18: t.665
141. Hopkins, H.C.F. (1994). The Indo-Pacific species of *Parkia* (Leguminosae – Mimosoideae). Kew Bull. 49(2) in press.
142. Hopkins, H.C.F. (1992). Two new subspecies of *Parkia* (Leguminosae – Mimosoideae) in Malesia. Blumea 37: 77–79.
143. Ohashi, H. & Sohma, K. (1970). A revision of the genus *Euchresta* (Leguminosae). J. Fac. Sci. Univ. Tokyo, sect. 3, 10: 210–235.
144. Chen, C., Sun, H. & Mizuno, M. (1992). On the genus *Euchresta* Benn. (Leguminosae) with 'Wallace's Line'. Acta Phytotax. Sin. 30: 43–56.
145. Brenan, J. & Brummitt R. (1965). The variation of *Dicrostachys cinerea* (L.) Wight & Arn. Bol. Soc. Brot., ser.2, 39: 61–115.
146. Barneby, R.C. (1991). Sensitivae Censitae. A description of the genus *Mimosa* Linnaeus (Mimosaceae) in the New World. Mem. New York Bot. Gard. 65: 1–835.
147. Chen Te-Chao (1988). Flora Reipublicae Popularis Sinicae 39. (Leguminosae (1)).
148. Larsen, K. & Larsen, S.S. (1993). Pers. comm.
149. Larsen, K. (1993). Note on the nomenclature of *Cassieae* (*Leguminosae* – Caesalpinioideae) in Malaysia. Nordic J. Bot. 13: 403–404.
150. Larsen, K. & Larsen, S.S. (1991). Notes on the genus *Bauhinia* (Leguminosae – Caesalpinioideae) in SE Asia. Nordic J. Bot. 11: 629–634.
151. Nielsen, I. (1993). Pers. comm.

152. Ohashi, H. (1982). The taxonomic position of the genus *Murtonia* (Leguminosae). J. Jap. Bot. 57: 225–231.
153. Ohashi, H., Tateishi, Y., Nemoto, T. & Endo, Y. (1988). Taxonomic studies of the Leguminosae of Taiwan. III. Sci. Rep. Tohoku Univ. ser. 4, 39: 191–248.
154. Ohashi, H. (1991). Taxonomic studies in *Desmodium heterocarpon* (L.) DC. (Leguminosae). J. Jap. Bot. 66: 14–25.
155. Ohashi, H., Tateishi, Y., Nemoto, T. & Hoshi, H. (1991). Taxonomic studies of the Leguminosae of Taiwan. IV. Sci. Rep. Tohoku Univ. ser. 4, 40: 1- 37.
156. Ohashi, H. (1992). A new species of *Desmodium* (Leguminosae) from Laos. J. Jap. Bot. 67: 320–323.
157. Ohashi, H. (1982). The taxonomic position of *Tadehagi rodgeri* (Leguminosae - Desmodieae). J. Jap. Bot. 57: 264–268.
158. Ohashi, H. (1993). Pers. comm.
159. Schrire, B.D. (1993). Pers. comm.
160. Lock, J.M. (1993). Pers. obs.
161. Gillett, J.B. (1963). *Sesbania* in Africa (excluding Madagascar) and southern Arabia. Kew Bull. 17: 91–159.
162. Verdcourt, B. (1987). Three corrections to the Flora of Tropical East Africa. Kew Bull. 42: 657–660.
163. Sauer, J. (1964). Revision of Canavalia. Brittonia 16: 106–181.
164. Verdcourt, B. (1993). Pers. comm.
165. van der Maesen, L.J.G. (1993). Pers. comm.
166. Maréchal, R. (1993). Pers. comm.
167. Greuter, W., Burdet, H. & Long, G. (1989). Med-Checklist Vol. 4. Dicotyledones (Lauraceae - Rhamnaceae). Geneva: Conservatoire et Jardin botaniques.
168. Ohashi, H. (1975). Nomenclatural changes in several Himalayan legumes. J. Jap. Bot. 50: 305–309.
169. Polhill, R.M. (1982). *Crotalaria* in Africa and Madagascar. Rotterdam: A.A. Balkema.
170 Lock, J.M. (1989). Legumes of Africa - a check-list. Royal Botanic Gardens, Kew.
171 Schnell, R. (1962). Remarques préliminaries sur quelques problèmes phytogéographiques du Sudest asiatique. Rev. Gén. Bot. 69: 301–366.
172 Barneby, R.C. (1991). Sensitivae Censitae. A Description of the Genus *Mimosa* Linnaeus (Mimosaceae) in the New World. Mem. New York Bot. Gard. 65. Pp. 835.
173 Tsoong Pu-chin & Ma Chi-yun (1981). A study on the genus *Sophora* L. Acta Phytotax. Sin. 19: 1–26 & 143–167.

INDEX